TJ
163.3
S67

THE
ENERGY-SAVING
GUIDEBOOK

by
Dr. George S. Springer, Professor, Mechanical Engineering
and
Dr. Gene E. Smith, Professor, Mechanical Engineering
College of Engineering, The University of Michigan

a **TECHNOMIC**® publication
TECHNOMIC Publishing Co., Inc.
265 W. State St., Westport, Conn. 06880

22841

THE
ENERGY-SAVING
GUIDEBOOK

©TECHNOMIC Publishing Co., Inc. 1974
265 W. State St., Westport, Conn. 06880

a **TECHNOMIC**® publication

Printed in U.S.A.
Library of Congress Card No. 74-81585
Standard Book No. 0-87762-147-0

CONTENTS

INTRODUCTION

The energy crisis is finally upon us. For awhile it was an abstract thing, more of a threat than a real problem, but no longer. Now nearly all of us feel the effects of the shortage in one form or the other. We can't keep our homes at the usual comfortable temperatures; our lights and appliances are affected by power black-outs and brown-outs; we've had to cut down on our driving; long distance travel has become difficult with closed gas stations and reduced airline flights. More disturbingly, thousands of people have been laid off because of the energy shortage.

The odds are if someone were to ask you "Are you worried about the energy shortage?" you would reply, "yes." Yet if you're like most other people, you probably still wonder what you can do about the problem. You realize you'll have to make sacrifices, but what can you do besides turning down the thermostat a few degrees and driving a few miles less?

Actually, there are many steps you can take to conserve energy, without having to give up your hard-earned comforts, and without having to alter your life-style significantly. Some energy saving steps are simple and require nothing more than a slight change in your habits. Others may require maintenance measures for your home and car which in all likelihood should have been done regardless of the energy situation. Still others may necessitate minor alterations on your home. But although various steps require varying degrees of effort on your part, they all have one thing in common: they're all simple and inexpensive. Even those methods which require a little investment save money in the long run. Their cost is recovered quickly by the money you save on heating, utility and repair bills.

The energy-saving measures you adopt can bring you hundreds of dollars in savings and a great feeling of satisfaction, and for most people that should be enough. But even if you can afford the increasingly high price of gasoline and electricity, you still have a stake in energy conservation. Right now energy is still a relatively cheap commodity, even when it's in short supply. When it runs out, though, no amount of money will be able to buy it. All the cash in the world won't do you much good at a closed gas station, or keep your lights burning during a black-out. The energy you help save today may be the energy you need tomorrow, and it's in your own interest to conserve it.

In this book we show you how to save energy at home, on the road, at work, and in your leisure time. Don't worry, though, if you can't adopt every method we suggest. Some measures can be used across the country: some are more useful in the North, some in the South, some in rural areas, and some in cities. Adopt those methods which are the most suited to you.

Another word of caution: your savings may not be exactly those suggested by the various charts and tables in the book. These figures are based on average conditions, and your conditions are likely to be a little different from the average. You may save a little more than the charts suggest, or a little less. Still, the charts and tables show you the general direction to follow in order to save energy. Use them as general guides, and then experiment to find exactly which conditions are the best for you. You'll be surprised at how much energy and money you can save.

CHAPTER I

A FEW FACTS ABOUT ENERGY

The purpose of this book is to help you beat the energy crisis, not to analyze it for you. It's not that we're not interested in what caused the shortage, or in whether it's long-term or short-term. We just feel that whether it's natural or contrived, long-term or short-term, the energy crisis is going to cause a lot of people a lot of hardship, and it's going to cost all of us a lot of money. We want to tell you specific ways in which you can reduce that cost and that hardship by as much as possible.

First, though, let's look at a few basic facts about energy — where it comes from, where it goes, and how we measure it. This information will help you understand some of the ways in which we save energy, and it may familiarize you with some of the terms you see used on TV and in the newspapers. Let's begin with where energy comes from.

THE SOURCES OF ENERGY

Energy, we all know, is the capacity to do work, and it can exist in a variety of forms: mechanical energy, chemical energy, nuclear energy, etc. Most of the energy we use in the United States comes from three types of chemical energy — namely, coal, oil, and natural gas. These three natural resources are called *primary energy resources* and together account for 98 percent of the energy we use. Oil and its derivatives — gasoline, heating oil, diesel fuel, etc. — provide 44 percent of our energy; natural gas provides 33 percent of it, and coal provides 21 percent. The remaining 2 percent comes from other resources such as nuclear energy and hydro power.

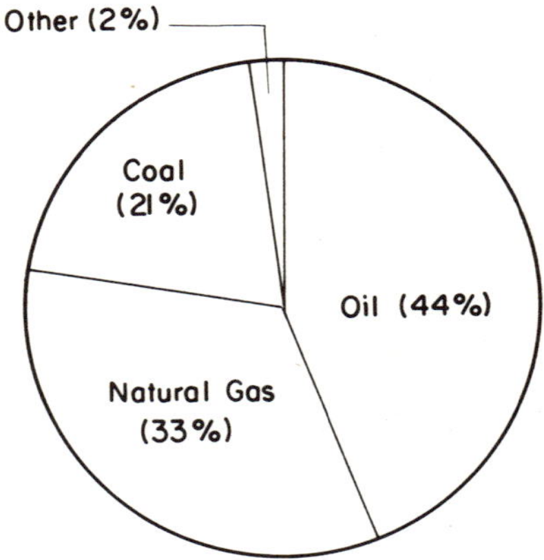

Secondary energy resources are different from primary resources in that they are produced from primary resources. Usually the process makes the energy easier to use. In this country, electricity is the main secondary energy resource, and it can be produced from any of the primary resources. Most of our electricity is generated from coal, but in recent years there's been a shift towards using more oil and gas in generating plants because they pollute less than coal.

For our purposes, what's interesting about the distinction between primary and secondary resources is the available energy that's lost in converting a primary resource to a secondary resource. When coal, oil or gas is burned to produce electricity, for instance, only 35 percent of the energy in the primary resource is converted to electricity. Some of this is then lost in the wires before it gets to your home, so only 30 percent of the energy in the coal burned by the electric company reaches your home in usable form. We'll refer to this fact later when we look at the relative efficiencies of oil, gas, and electric appliances.

THE USES OF ENERGY

Since energy is simply the capacity to do work, there are almost an infinite number of ways it can be used. All we're interested in, though, are the broad categories of energy use in this country, so we can form some idea of where our energy is going.

In the United States, about 43 percent of the energy consumed powers various industrial processes. Home heating accounts for 11 percent of the total energy used; commercial heating accounts for 7 percent; automobile transportation accounts for 13 percent, and public transport for 12 percent. Then we have hot water heaters (4 percent), air conditioning (2.5 percent), refrigeration (2 percent), lighting (1.5 percent), cooking (1.5 percent), and all other uses (2.5 percent).

The point of all this is that a significant proportion of the energy used in this country is under the control of ordinary citizens. We can't do too much about the 43 percent used in industrial processes, true, but by being careful in the way we drive, heat our homes, cook our food, and perform other everyday activities, we can have a big impact on the energy shortage.

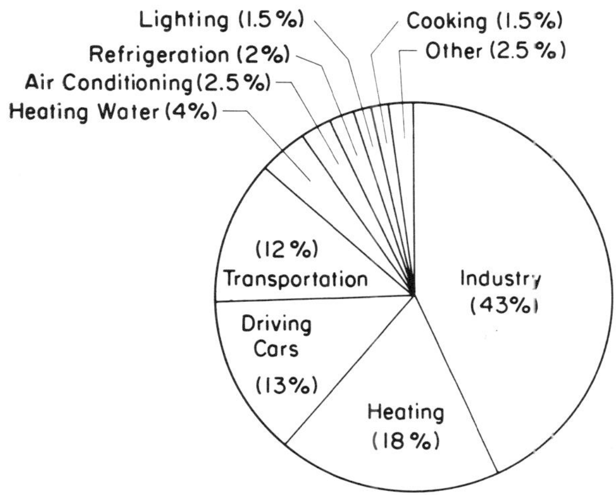

The Uses of Energy

Whenever energy is used, of course, a certain amount of it is wasted. It isn't actually destroyed, but it is changed into an unusable form, and for our purposes it might as well be destroyed. The fraction of total energy used that a machine converts into usable work is the machine's efficiency, so that an automobile engine with an efficiency of 25 percent will convert 25 percent of the energy in the gas it burns into mechanical energy to move the car. The other 75 percent will be wasted.

Naturally, you want to use the most efficient engines and appliances possible in order to save energy. If you're buying a furnace and can get either a gas- or an oil-burning model, for instance, choose the gas model. Its efficiency is probably around 70 or 80 percent, while the oil furnace will turn only 60 or 65 percent of the energy in the oil into heat you can use to heat your home. If gas isn't available in your area, of course, you don't have any choice.

When you're thinking about efficiencies, remember what we just said about primary and secondary energy resources. It might seem at first that an electric clothes dryer (efficiency equal to almost 100 percent) will waste less fuel than a gas-operated dryer (efficiency equal to 80 percent). Remember, though, that the efficiency of the electricity-generating and transmitting system is only 30 percent, so the gas dryer actually uses less fuel than the electric one.

How much energy do you use personally? To answer this question, you ought to check your electricity meter and gas meter every so often, or see how many gallons of fuel oil you use in a month. You also need to record how many gallons of gasoline you put into your car. Your electricity meter records the number of kilowatt-hours of electricity you use, giving you a measure of electrical energy. Your gas meter records the volume of gas you use in cubic feet, and you can convert this to energy by figuring that each cubic foot of gas contains about 1000 British Thermal Units (Btus) of energy. (This figure, the "heating value" of the gas, may vary somewhat in different parts of the country). Heating oil contains about 145,000 Btus per gallon, and gasoline about 113,000 Btus per gallon.

If you're the "average American homeowner" (which you're probably not, since no one is "average"), you drive your car 12,000 miles per year. At 15 miles per gallon of gasoline, you use 800 gallons or 90 million Btus of energy for driving every year. This is equivalent to about 26,500 kilowatt-hours of energy. By contrast, you use only 8,000 kilowatt-hours of electricity per year, but since the electric company converts coal to electricity with only a 30 percent efficiency, 26,700 kilowatt hours of energy are needed to produce the 8,000 kilowatt-hours you use. If you have a gas furnace you use about 150 million Btus to heat your home every year — another 44,000 kilowatt-hours. Your total energy consumption, then, is almost 100,000 kilowatt-hours per year.

What does 100,000 kilowatt-hours mean though? To answer that, let's look at how energy is measured.

THE MEASURES OF ENERGY

There aren't really as many ways to measure energy as there are to use it, but sometimes it seems that way. An auto mechanic uses one measure of energy, a nutritionist another, an atomic physicist still another, and so on. For our purposes,

Electric Meter

Reading: 67,412 Kilowatt-hours

Gas Meter

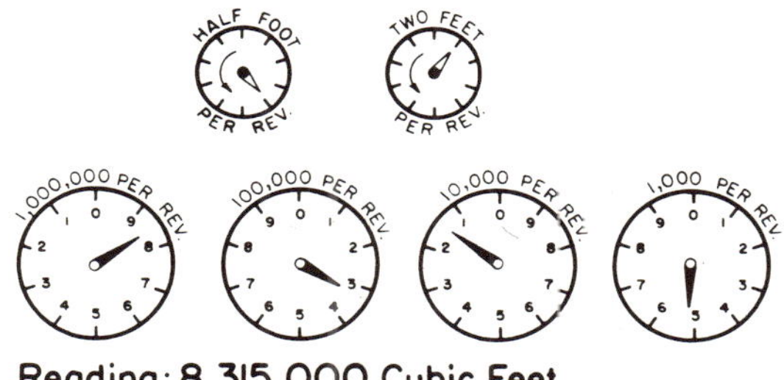

Reading: 8,315,000 Cubic Feet

How to Read Your Own Meters

though, there are three main measures of energy: the British Thermal Unit, the watt-hour, and the kilowatt-hour. A table showing how to convert from one set of units to another is given in the Appendix.

The British Thermal Unit, or Btu, is generally the unit we use to express the energy content or "heating value" of various fuels. One Btu is the amount of energy required to raise the temperature of one pound of water by one degree Fahrenheit. It's also about the amount of energy you release by burning a match. As noted above, one cubic foot of natural gas yields approximately 1000 Btus when it's burned in your furnace. The number of Btus an appliance uses over a time is expressed, not surprisingly, in Btus per hour.

A watt-hour is the amount of electricity a one-watt appliance uses in an hour. A kilowatt-hour is equal to 1,000 watt-hours. An appliance's wattage, you probably remember, is equal to the voltage it operates on multiplied by the current needed to run it. A 120-volt appliance which uses 10 amperes of current, for instance, has a wattage of 1200 watts.

The watt-hours or kilowatt hours that an appliance uses will depend upon three things: its voltage, its amperage, and the amount of time it's operating. You can control the amperage and the operating time to some degree by buying appliances which run on a low current and by operating them infrequently, but you can't control voltage. Voltage is set by the electric company. In this country voltage is usually maintained at 120 volts, and power lines for some major appliances have a voltage of 240 volts. When the electric company can't keep up with the demand for electricity, it reduces the voltage of the system, thereby providing you with less electricity. This is what we know as a "brown-out".

CHAPTER II

HEATING YOUR HOME

Heating the home represents the largest single residential use of energy in the United States, accounting for approximately 11 percent of the total energy used in this country. It is one of the obvious areas you should consider then, when you explore ways to reduce your energy consumption.

The amount of energy required to heat your home depends upon the amount of heat which is lost through the walls, ceilings, floors, windows and doors. This "heat loss" is the result of temperature differences between the inside and outside of your home as well as air leakage through various openings. More generally, it depends upon the house size, the type and quality of construction, the amount of insulation present, and the severity (in terms of temperature levels and wind velocity) of winters in your area. For an average home in the midwest, total heat loss during a winter heating season is typically about 120 million Btus. To provide this much energy you have to burn 172,000 cubic feet of gas (costing $189.20 at $0.11 per 100 cubic feet) or 1450 gallons of fuel oil (costing $362.50 at $.25 per gallon), so any improvement in your heating system or decrease in heat loss from your home represents significant savings on your heating bill. A 10 percent improvement in the system, for instance, will save you $19.00 per winter if you heat with gas, or $36.00 per winter if you use oil.

In this chapter you will discover many ways to reduce the heat loss from your home and improve the operating efficiency of your heating system. Some of these methods will save you a lot of fuel all by themselves. Others may reduce your energy consumption by only a little, but when all those little savings are added together they may have a big effect on your heating bill. Again some of the suggestions may require some sacrifices in your personal comfort (for instance, lower temperatures) — others, however, will not. As you read through this chapter make mental notes or, better yet, paper and pencil notes, of the methods which could apply to your home — then see how many of them you can put into practice. To help you with this, we've included a check list of our own at the end of the chapter, listing ways in which you can save both energy and money while heating your home.

FURNACE

The job of your furnace is to replace the heat lost from your house in order to maintain the temperature level you desire. There are three main ways to do this. You may have a fan in your house which blows air through the furnace and then returns it to the rooms through an air distribution system. On the other hand, your furnace may heat hot water which circulates through radiators in the house, in turn heating the air in the different rooms. Or your furnace may produce steam which circulates through your radiators. These three methods of heating the home are referred to as hot air, hot water and steam heating systems, respectively. Regardless of which method you use, the chances are that either a gas or oil flame is used to heat the air, water or steam involved.

In some homes, however, the energy source is electricity, or (in some older installations) it may even be coal. If your home is heated by electricity, it's quite likely this is done by resistance heaters located throughout your house, in which case you have no furnace. If this is the case, many suggestions made in this section will not apply and those that do apply will be obvious. But if your home has a coal-fired furnace most of the suggestions made here will apply. No detailed discussion of coal-fired furnaces will be presented, though; our emphasis will be on gas and oil fired furnaces.

The efficiency of your furnace is defined as the percentage of energy available in the fuel (commonly called the heating value) which is actually transferred to the air (or water or steam) passing through the furnace. Typical efficiency values range from 40 to 80 percent, with the energy which is not transferred to the air being lost out the chimney. While you can't make your furnace 100 percent efficient, there are several things you can do to make the efficiency as high as possible. First, it's very important that the proper amount of air be mixed with the fuel. If too little air is supplied, the fuel won't burn completely and unburned fuel will be lost out the chimney. This pollutes the air, wastes energy, and wastes your money. On the other hand, if too much air is supplied to the furnace, part of the fuel is used to heat the extra air which is then lost out the chimney. Setting the ratio of air and fuel in your burner is called "flame adjustment" and is generally done by the manufacturer at the factory or during installation in your home. It might save you money, though, to have a competent heating contractor check your furnace flame adjustment periodically to assure efficient combustion. The contractor's fee can easily be recovered in fuel savings, and flame adjustment isn't something you should try to set on your own.

Cleaning your furnace every so often will also improve its efficiency. The heat produced by the flame in the furnace is transferred to the air, water or steam across a series of metal passages. If your furnace flame adjustment is faulty soot may accumulate on the combustion side of these passages, acting as insulation and significantly reducing the efficiency of the heat exchange process. The effects on your heating bill can be considerable. A layer of soot 1/8 of an inch thick can increase fuel consumption 10 percent, costing you up to $40 or $50 each winter. Soot build-up is more common in oil-burning units than in gas-burning units because gas burns more cleanly. Whether your furnace operates on gas or oil, though, it's still a good idea to have it cleaned regularly. This can be done at the same time you have the flame adjustment checked.

In electrically heated homes there's no furnace or furnace chimney to worry about, so heat can't be lost up the chimney. For this reason electric heat is said to be 100 percent efficient. But in spite of this high efficiency, electric heat uses more of our primary fuels (gas, oil and coal) than comparable oil or gas furnaces do. The losses in coverting primary fuels to electricity at the generating station, in transmitting the electricity along power lines, are such that only about 30 percent of the energy available in the fuel is delivered to your home as electricity. Thus your oil furnace may have an efficiency of 60 percent and use 1450 gallons of fuel during the heating season, but if you went to electric heat the electric company would have to burn 2900 gallons of fuel each winter to heat your home. Clearly, electric heat is no way to conserve energy. But back to your oil or gas furnace. If you have a hot air system, make sure the filters are cleaned or replaced regularly. A dirty filter will not only restrict air flow and reduce furnace efficiency — it will also increase blower operating time and add to your electricity bill. If your filters are extremely dirty, air flow may become so restricted that your furnace will be unable to heat your home adequately.

Other components of the air handling and distribution system of your furnace must also be properly maintained if you want efficient operation. The blower and blower motor should be lubricated regularly (usually once a year) unless your unit has sealed bearings. If you have sealed bearings, they should be obvious — there won't be any lubrication ports or openings. If you don't have sealed bearings, follow the manufacturer's recommendations regarding lubrication.

Also, the blower itself should be cleaned for more efficient air handling. This can usually be done with a household vacuum cleaner, and it shouldn't be too much of a job. If your air filters are properly maintained all air handled by the blower will have been filtered first, and the blower won't have much dust accumulated on it. If your furnace uses an electronic filter, be certain it is properly maintained and operating correctly. Malfunction of this unit allows house dust to pass directly through the blower, causing it to build up on the blower and in the air ducts leading to the rooms.

While we're on the subject, you should check the air ducts which circulate air to and from the rooms in your house to be sure all joints and connections are tight. Leaks in the supply ducts carrying warm air to the rooms will cause heat to be lost to your basement or to the crawl spaces under your home. Such leaks are not necessarily bad since they will provide some heating for the floors. However, as a general rule, all such leaks should be eliminated. Leaks in the return ducts carrying air back to the furnace from the rooms also cause less efficient operation and should be avoided. You can purchase special duct sealing tape at hardware, lumber and heating supply stores and seal most joints and connections yourself.

One more tip for owners of hot air systems: make sure that the hot and cold air registers in your home are not obstructed by furniture, drapes, rugs or other furnishings. These obstructions lead to inefficient air handling and may also cause some parts of your home to be cooler than others.

Now let's suppose your home is heated with hot water or steam. The first thing you should do is make sure there are no leaks in the pipelines and convectors or radiators. Such leaks, which are usually quite obvious, cause water or steam to be heated but not used for your purposes. They also force you to add water to the system, and unless your water is specially treated this make-up water will contain minerals which will be deposited on the inside of the pipelines and radiators. The deposits in turn decrease the heat transfer efficiency in your furnace and your radiators, causing fuel to be wasted. For this reason it's unwise to replace the water in your hot water or steam heating system unless you're sure the water you put in contains few minerals. A heating contractor can tell you whether the water in your area is high in minerals or not.

Another thing to watch for in a water or steam heating system is air which may become trapped in the lines and radiators. This trapped air can block your water flow or cut down on the effectiveness of your radiators. A properly installed system will have air bleed valves at certain points, and you should check these valves regularly to make sure there's no air in the system. If you feel uncertain about making such checks yourself, ask a qualified heating contractor to do it for you. Also, if your home is heated by steam, make sure that all return lines in the system have the proper slope, so the condensed steam can return easily to the furnace. Low spots or incorrect slopes in your return lines may restrict the steam flow in parts of your heating system.

You can carry out many of the above suggestions yourself, but some heating maintenance should be done only by a qualified professional. A heating contractor can check most of these items in one visit, and the money you'll save on fuel will more than pay for his service charge. These maintenance procedures will also contribute to a longer life for your heating system.

INSULATION

The previous section dealt with ways in which you can increase the efficiency of your furnace. This and following sections will tell you how to save energy by reducing the heat loss from your home, so your furnace won't have to provide so much heat in the first place.

Adding insulation to your home is one of the most effective ways of reducing heat loss. In most homes being built today insulation is a standard feature, but you may want to improve upon what the builder installs. In an older home, the addition of an inch or two of insulation may lead to truly amazing reductions in your heating bill.

Suppose, for instance, that you live in an uninsulated one-story house in the midwest. Adding one inch of insulation to the ceiling will cut your total heat loss by 33 percent, and if you add that one inch of insulation to the walls as well, your heat loss drops by a total of 46 percent. Although thicker insulation will give you additional savings, the first inch is what really counts. Four inches of insulation on your ceiling cuts the heat loss of a previously uninsulated house by 37 percent; four inches on the ceiling and one inch in the walls causes it to drop by 53 percent.

These figures depend upon the style and location of the home and therefore, may not describe your heating situation exactly. The figures do point to some general trends, though, which will apply to almost everyone. First, the most effective place to install insulation is in your ceiling. This is where much of the heat from your home escapes. Second, the first inch of insulation is the most effective. More insulation will give you more reduction in heat loss, but the effect is less for each additional inch. Finally, even though insulating your ceiling has the most effect on heat loss, you can save a significant amount of energy by insulating your walls, too.

How much money will insulation save you? Well, suppose you own an uninsulated two-story home (1250 square feet of living space) built before 1940. Putting six inches of insulation on your ceiling and none in your walls will reduce your annual heating bill by 17 percent, and your savings will pay for the insulation in a little more than two years. Or suppose you own a one-story home (1100 square feet) built after 1940, and it already has 1½ inches of insulation on the ceiling and walls. Adding 4½ inches of insulation to the ceiling will cut your heating bill by 10 percent, and the resulting savings will pay for the insulation in a little more than six years.

Remember, though: since you get a diminishing amount of savings for each extra inch of insulation you buy, you don't want to smother your home with the stuff. Instead, you want to balance the cost of each inch of insulation against the amount

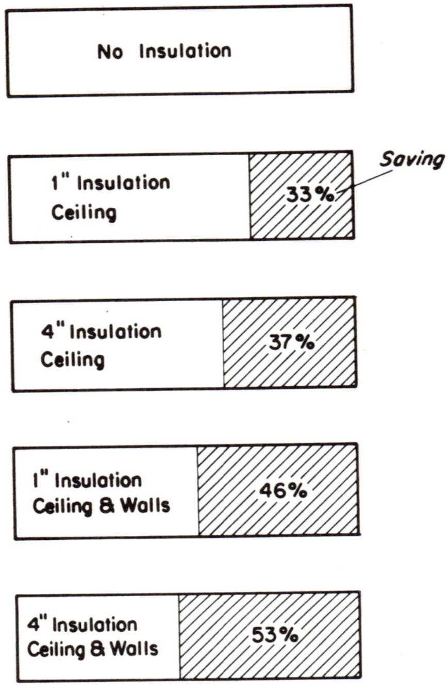

Effect of Insulation on Heat Losses from a Single Family House.

of money it will save you on your heating bill. You're going to buy each extra inch if, and only if, it makes you a profit. Or, in fancier language, you're going to buy the "economic optimum" of insulation for your home.

This economic optimum is not necessarily what the contractor put into your house, even if the house had to meet FHA standards. The standards, after all, only define the minimum insulation the builder can get away with and still offer you an FHA mortgage, and you may want to add more to get up to the optimum.

Given that you want to install the economically optimum amount of insulation, though, what is this optimum for your home? That depends on the climate in your area, your house design, heating habits, heating fuel costs, and insulation costs. You really ought to ask a reliable heating contractor to show you the tradeoffs involved. Generally, though, you should have at least two inches of insulation on the outer walls of your house and four inches on your ceiling or in your attic. In a very cold climate, you might want four inches on the wall and six inches on the ceiling. Unless your house is still under construction, you probably can't do anything about your walls, but you may be able to add insulation to your ceiling by installing it through your attic.

If you're planning to install the insulation yourself, be sure first to consult with an insulation contractor or building supply store. Find out what kinds of insulation are available, what's recommended for your area, and what exactly you should do to install it. Be especially careful to install the vapor barrier properly. The vapor barrier keeps moisture inside your house from seeping through walls and ceilings into the insulation and reducing its effectiveness.

INSULATING GLASS AND STORM WINDOWS

Our ancestors learned a long time ago that a piece of glass placed in an opening in the cabin wall helped make the cabin more comfortable. Not only did this keep wind from blowing through the cabin, but it also provided some insulation or resistance to heat loss through the window opening. What was not obvious to our ancestors and is not obvious to most people today is that the insulating effect is not due only to the glass. Still air is good insulation, and the major insulating effect of a glass window comes from the relatively calm layer of air on either side of the glass, not from the glass itself. Thus the thickness of the glass is of no major importance in reducing heat loss. On very windy days the air on the outside of the glass no longer remains calm, and part of the insulating effect of the glass window is destroyed. A window screen, however, will help to trap the air even on windy days, so leaving your screens up all winter may actually help insulate your windows. (This is nice to know around Christmas time when you're trying to tell your spouse why you didn't remove the screens in the fall.)

Even better than the screen, though, is adding an additional layer of glass to your window, thus creating a dead air space between the two panes of the glass. This additional layer of glass may be in the form of a storm window or in the form of an insulating glass. A storm window usually comes in a separate frame and is easily placed over an existing window; an insulating glass, on the other hand, consists of two glass panes separated by a metal band which completely seals the panes around their edges. Because of its construction, the insulating glass is difficult to use in place of an existing single pane window and usually requires a new window installation. A storm window, then, is what you'll want for existing window installations. A well-fitted storm window combined with a regular window will give you about the same amount of insulation as an insulating glass.

Storm windows can lead to moisture problems under certain circumstances. In the winter, the air in your home generally contains more moisture than the outside air. Because windows are not completely air tight, this moist inside air may leak into the air space between your window and your storm window and condense on the inside of the storm pane. For this reason, storm windows contain ventilation holes to let the moisture excape, thereby decreasing the moisture condensation problem. Resist the temptation to plug up these holes. Covering the holes will not improve the insulating value of your storm window, but it will steam up your windows. Remember that the insulating effect of the storm comes from the trapped air between the panes, and leaving the ventilation holes open won't disturb this air layer. If you have an insulating glass you won't have to worry about condensation — the air-tight seal keeps it from occurring.

How much money will storm windows or insulating glasses save you? Well, if you add a storm window to a window which has a single layer of glass, or if you replace this window with an insulating glass, you can reduce the heat loss through the window by about 50 percent. In an average home with no storm windows or insulating glasses, 20 percent of the heat loss occurs through the windows, so reducing this loss by 50 percent should reduce your heating bill by 10 percent. If you had three layers of glass per window — either a storm window and a two-pane insulating glass on each window, or a three-pane insulating glass you might increase your overall savings to 13 or 14 percent.

The total reduction in heat loss to be gained by installing storm windows or insulating glass in your home, of course, depends upon the number and size of your windows. Adding storm windows to existing single pane windows or using two-pane insulating glass will almost always save you money. Three-pane installations, though, are not always a good bargain. Before you buy them, check to be sure that you can save enough on heating bills to pay for the cost of installation. Unless you have a very large window surface in your home, you'll probably find that three-pane windows are a waste of money.

One cheap substitute for that third layer of glass — or even for the second layer of glass — is covering your window frames with polyethylene plastic sheet. This can be done by tacking the plastic to the outside of the window frames, although in some areas you may find that the polyethylene flaps against your house or even rips in the wind. Fastening the polyethylene to the inside of the windows eliminates these undesirable side effects. If the plastic is attached neatly, either inside or outside, it won't be too noticeable — especially if your curtains or drapes are closed.

Some additional comments on windows: if you open your windows in winter in order to air out your house, it's better to open many windows for a short time rather than a few windows for a longer time. Opening several windows for a short time lowers the air temperature in the room, but in a short time your walls and furnishings can't cool down, and they take more energy than the air in your house to warm up again. (Raising air temperature is easy, in fact — to raise the temperature of all the air in a 1400 square foot house by one degree you need 200 Btus, or enough energy to boil half a cup of water). You increase your energy saving even more if you only open those windows during the warmest part of the day.

Finally, if you have an air conditioner in one of your windows, remove it in the winter time. If you can't remove it, at least insulate it very well and cover it with a weatherproof covering.

WEATHER STRIPPING AND CAULKING

Even in the best-built houses, cracks form around door and window frames after a period of time, allowing warm air to escape from and cold air to enter the home. These air leaks may account for 10 to 35 percent of the heat loss from an average well-insulated home, so it's worth your while to eliminate them.

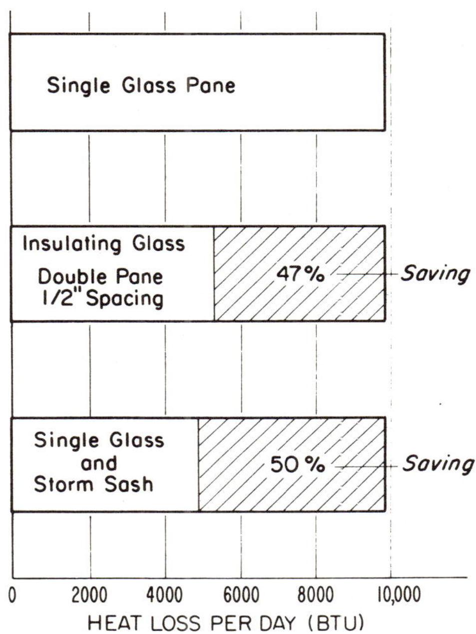

Typical Heat Loss per Day from a 10 Square Foot Window.

You can reduce air leaks by applying weather stripping and caulking to your home. To do this properly, first go around the outside of your house checking for possible leaks. Pay especially close attention to the joints formed where door and window frames come up against the siding on your house; these joints should be sealed by applying caulking all around the edges of the frames. Also check all windows to make certain that the glass panes are tight in their individual frames: if they're loose, apply caulking or putty to seal the cracks. Do you have a mail slot in the door which you don't use? Seal it off with tape. Or if you do use the mail slot, consider putting a closed box over it to receive your mail. It saves energy and will keep your letters from falling on the floor. The same principle goes for milk delivery chutes — if they're not being used, they should be filled with insulation and the doors sealed with tape. Air leaks in the siding of your house should also be sealed.

The sides of your house aren't the only places where heat can escape. If you have crawl spaces under your house, you will have air vents in your outer walls under the living area to prevent moisture accumulation and the resultant rotting of building materials. In the winter time, the moisture content of the air is low and you should partially seal these vents. (Be sure you open them again in the summer, though, when moisture does build up.) You should also check the moving portions of your windows: if they don't have weather stripping already, you should get some. The metal kind is best, since it will last. Felt or rubber weather stripping will wear out much sooner and should only be used for temporary jobs. You should also put weather stripping around the edges of your outside doors. Check at your favorite hardware store or lumber yard to find out what's available and how you should install it.

Now check the inside of your home for possible air leaks. Make sure the interior wood trim around outside doors and windows is tight against the walls and frames, and if there are any openings, seal them with putty. Also check to see that you have a good seal between the bottom of all outside doors and the step or threshold. If the seal is absent or defective, install one of many which are available at the hardware store or lumber company, or at least roll up a small rug and lay it across the bottom of the door.

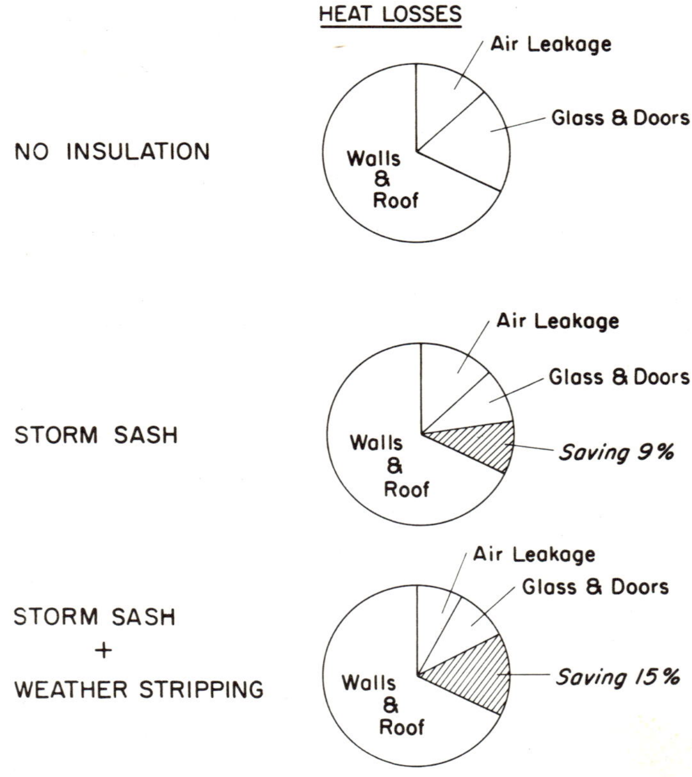

Effects of Weather Stripping and Stormwindows on
Heat Losses from a Single Family House.

Continuing your inspection, check all windows cn the inside to make sure they're tight. If you don't plan to open and close a window frequently you can seal around the edge of it with tape, although you may not want to do this for every window. If you have sliding windows, be sure to check the joint where the two window-halves join in the middle of the window. And finally if you have a door or hatch leading to the attic, install weather stripping around the edges. It wouldn't hurt any to add insulation to the back side of the door, also.

The key to reducing air leaks in your home is to look diligently for all possible leaks inside and outside the house and then to apply caulking, weather stripping, putty and tape as needed. This won't only reduce your heating bill — it'll also eliminate those small drafts which can be so annoying and uncomfortable.

DRAPES AND SHADES

Drapes can help a lot in reducing window heat losses from your home. They have little insulating value in themselves, but they make a major contribution to energy conservation by trapping air next to the windows. In this way they're similar to an additional pane of glass, though not as effective. Since the main value of drapes is in trapping air, the particular fabric is not important as long as it's a fairly close weave. The thickness of the fabric is of no importance from an insulating standpoint.

The color of your draps does have some effect on transmitting the heat of the sun into your home. This solar heat is of lesser importance than the insulating effect of the air trapped by the drapes, so you won't want to change an otherwise acceptable decorating scheme just to increase solar heating. If you're changing your drapes anyway, you might remember that dark colored drapes will absorb sunlight, while lighter drapes will reflect it back outside again. Thermally insulated drapes are lined on the window side with light colored material to reflect the sun's rays.

Ideally, then, you want dark drapes in the winter and light or thermally insulated drapes in the summer. Since you probably don't want to buy different drapes for different seasons, you might be wise to buy light-colored or thermally insulated drapes. They give you insulation in the winter because of the trapped air, and since they reflect the sun's rays they also help keep you cool in summer. If you're really set on capturing that solar heat in the winter, though, you need dark drapes. You can make these suitable for the summer by lining them on the window side with a removable, reflectorized insulating liner that you can buy in most department stores.

On bright, sunny days you'll generally want your drapes open to let in the sunlight. After all, the sun's warmth is free, so why not take advantage of it? But if it's very cold or windy out, close the drapes even though the sun is shining: again, the sun's heat is less important than that insulating layer of trapped air during such conditions. Drapes should also be closed at night and on cloudy days when there's no sunlight to take advantage of.

You've undoubtedly experienced the warming effect of a bright fire burning in a fireplace when the fire makes you feel warm or perhaps even hot although the room air may be cool. This effect is caused by the heat radiation your body receives from the fire. This radiation is not affected by the room air temperature. The opposite effect occurs when your body radiates heat through uncovered windows to the cold outside air, producing a "cold radiation effect." Under such circumstances, increasing the air temperature by turning up the furnace won't keep you from feeling cold because the radiation will continue. Covering the windows with drapes will reduce the radiation and you'll feel warmer.

Window shades have about the same effect as drapes on heat loss, although since they're usually much closer to the window, they're not quite as effective in trapping air. Still, they should be raised and drawn just as you would open and close drapes. Drawn shades also help reduce the cold radiation effect through glass.

Venetian blinds are like drapes and shades, except they have an additional advantage. The angle the slats form with the window can be varied, and you can set this angle so the blinds trap some air against the window but still let sunlight into the room. You should close venetian blinds entirely, though, at night and on cloudy days.

As we all know, the sun is low in the sky in the winter and high in the summer, and you can save energy with a special screen that takes this into account. The screen contains many small slats, similar to those in a venetian blind, set at an angle that lets sunlight through in the winter but **blocks** it in the summer. You might consider installing such screens on windows with a southern exposure.

Since the insulating effect of your drapes comes from the air trapped between them and your windows, you want to be careful not to disturb this air. If air registers or radiators are located in the space between the drapes and the windows, you should install deflectors to divert the warm air flow outside the drapes. Deflectors can be purchased at heating supply stores and certain hardware stores.

You can insulate the floors in your house by carpeting them, but it won't reduce your heating bill by very much. The heat lost through your floors is usually considerably less than that lost through the ceiling, walls, windows and doors, so the cost of carpeting can't be justified by its insulating value. Carpeting will make floors warmer, though. If you go barefoot a lot, or if you have children who play on the floor, carpeting will add to your family's comfort.

It's difficult to predict just how much heat you can save with drapes, shades, and carpeting. In an average home, your savings could range anywhere from 5 to 20 percent, and this should be enough to make you consider getting drapes as part of your energy saving program, if you don't have them already.

SETTING THE THERMOSTAT

The setting of your thermostat can have a very significant impact on your heating bill. The heat loss from your home is essentially a function of the difference between the inside and outside air temperatures, and by changing the inside temperature you change this difference. Consequently you change the amount of money you spend on heating.

Generally, changing your thermostat one degree will change your heating requirements by about 3 percent. If you normally keep your house thermostat at 70°F, changing this to 69°F will save you 3 percent on your heating bill; changing it to 68°F will save you 6 percent, and so on. Increasing it to 71°F will add 3 percent to your heating bill. Wearing a sweater which lets you stay comfortable while lowering your thermostat a degree or two, then, can save you a fair amount of money. Experimenting with the thermostat until you find the lowest comfortable setting is another good idea.

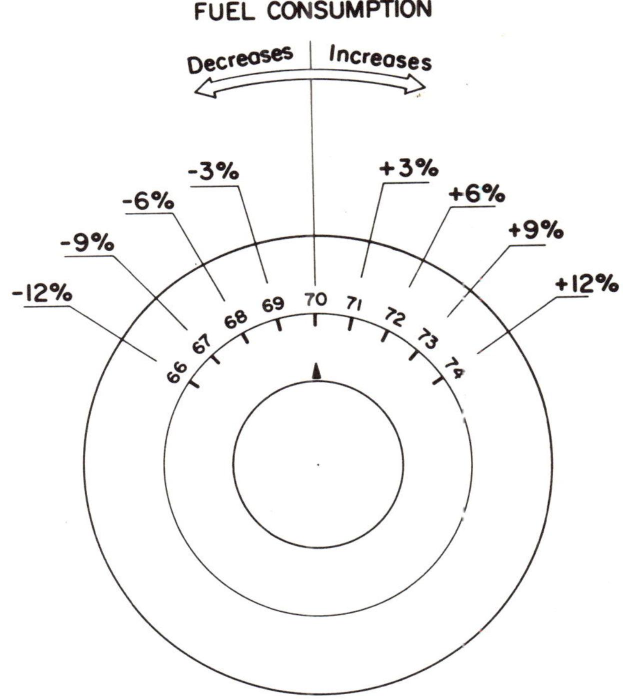

Effect of Thermostat Setting on Heating Fuel Consumption.

Once you've found this setting, leave the thermostat at that temperature. You may want to lower it a little at night, but only a little: at too low a setting your house structure and furnishing will cool off and then require large amounts of energy to reheat. This could reduce any fuel savings you achieve at the lower temperature. As a rule of thumb, you shouldn't lower the thermostat more than 10 percent below your daytime setting.

Your nighttime setting is appropriate during the day when everyone is away from your house, but if you're going to be away for more than a day, you can set the thermostat even lower. Be sure that this low setting doesn't shut off your heating system entirely, though. Also, when you return to your home, don't set the thermostat higher than normal in hopes that this will make your furnace work any faster. Your house will warm just as fast at the correct setting as it will at a higher setting.

You may find it convenient to buy a thermostat with a timer and dual settings. This allows you to select the desired daytime and nighttime settings and the times you want each to be in effect. The thermostat will then automatically lower the temperature in the evening after you have gone to bed and raise it again before you get up in the morning.

If you have more than one thermostat in your home, as might be the case if you have electric heat, you may not be able to operate them independently of one another. As a rule, thermostats in adjacent rooms should not be set more than 5°. apart. Any greater difference will tend to make the heating system in the warmer room try to heat the cooler room as well, and will lead to a waste of fuel.

However many thermostats you have, all of them will need their calibration checked from time to time. It's not uncommon for a thermostat setting to be a few degrees above or below the temperature which the setting actually produces. To check calibration, place a thermometer next to your thermostat and at several other places in your home to see what the temperature actually is. If you check the actual temperature against three of four thermostat settings, you can make a little table of thermostat settings versus actual temperatures and put it near your thermostat. The table will then tell you how to set your thermostat to get the room temperature you want. Incidentally, don't measure room temperatures by hanging the thermometer on an outside wall, because the wall will be a few degrees cooler than the air in the room.

Which brings up another point: the location of your thermostat affects the temperatures it records, so you don't want it in either an unusually warm or unusually cold place in your home. Placing it near the chimney, against an outside wall, near a radiator, or in direct sunlight, then, is a bad idea. The best place for your thermostat is on an interior wall which is near the center of the area you want to heat, and which doesn't expose the thermostat to sunlight from a window or hot air from a radiator or register.

Of course the builder of your house put your thermostat where it is, and there isn't much you can do to move it now. There are several ways, however, in which you can compensate for a bad location. A lower setting on a thermostat in a cold location, or a higher setting on one in a warm location, is probably the most obvious remedy. You can find the exact setting you want by using a room thermometer to determine what the temperature in your house is at different thermostat settings, then choosing the setting that gives you the desired temperature. If your thermostat is exposed to sunlight, you might also construct a simple shade in front of it. Or if your thermostat is located above a register or radiator, you might put a small deflector below the thermostat to keep warm air from passing directly over it. Don't place the deflector immediately below the thermostat because room air must be able to circulate around it and through it if it's going to control your heating system properly. For the same reason, don't wrap tape or other material around your thermostat. This also will obstruct air flow.

CONTROLLING HUMIDITY

There are a number of reasons why it's important to have the proper humidity in your home. Moisture in the air keeps you healthy by keeping the delicate membranes and linings in your nose and throat moist and by reducing the drying of your skin. It also prevents many of the electric shocks you can get from walking across a very dry carpet, and it reduces the drying out of your furniture, which can cause glued joints to loosen. Finally, moisture in the air will help you feel comfortable at lower temperature levels, thus helping to reduce your heating bill. If you feel comfortable in almost completely dry air at 72°F, you'll probably still feel comfortable if the relative humidity is raised to 30 percent and the temperature is lowered to 70°F. This 2° decrease in temperature will reduce your heating bill by about 6 percent. Normally, the relative humidity in your home should be kept at approximately 35 percent during the heating season, unless the outside air gets very cold.

There are several ways to introduce more moisture into your home. A considerable amount is produced by cooking and some also is given off by showering, bathing and laundering, but, you shouldn't depend upon these as your only source of humidity. If you have a hot air heating system, you should consider installing a humidifier in the furnace. This humidifier should be maintained according to the manufacturer's recommendations, of course, and you should check it regularly. If the maintenance on the humidifier isn't good, the unit can become clogged with mineral deposits, and it won't work efficiently.

If for some reason you can't install a humidifier on your furnace, at least place pans of water on or near the registers. (In a hot water or steam system, the pans go on the radiators, but the idea's the same.) The pans should be shallow and wide rather than deep and narrow: it's the exposed area of water that counts, not the volume. Just be careful to put the pans where they won't obstruct air flow from the registers and radiators too much.

For a little extra money, you can also purchase room humidifiers. These can be used with any heating system and are designed to blend in with most home decors. You should put them where they won't blow directly on people, since the air coming from them will be damp and cool until it mixes with the heated air in the room. Incidentally, common vaporizers are not effective humidifiers. They're very effective for sick people, but they're not designed well for humidifying a home.

A less obvious way of humidifying your home is by using the exhaust air from your clothes dryer. This air is not only damp, but also warm. If your dryer is on the first floor of your home you might consider piping its exhaust into your living room, dining room, or kitchen. First, however, place a mesh sock of some sort over the exhaust duct so lint won't be blown into the air. Clean the lint from the sock regularly. If your dryer is in the basement you can still direct its exhaust inside the house, but it won't have as significant an effect on humidity in your living areas. If you use a gas dryer check to see that exhausting the air into your home will not introduce harmful combustion gases. Of course the dryer should be vented to the outside in the summer time.

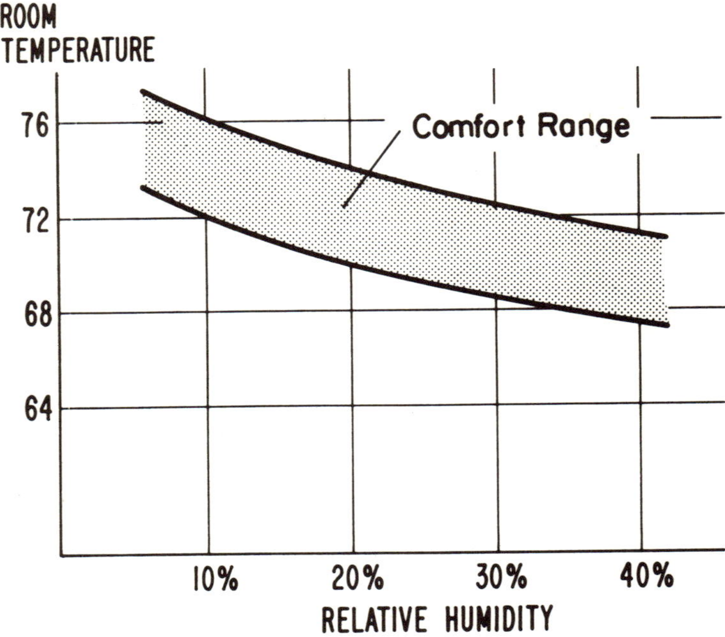

Recommended Room Temperatures and Humidity in the Winter.

As the temperature of air decreases the amount of moisture it can hold also decreases, so if the humidity in your home is too high water may condense on the relatively cold surfaces of your windows. To avoid this, you may have to decrease the humidity on very cold days. If you keep relative humidity at 40 percent and room temperature at 70°F, water will begin to condense on your windows when the glass cools down to 45°F. If you lower the humidity to 30 percent, condensation won't occur until the glass temperature falls to 37°F. The amount of moisture you can have in the air before condensation occurs depends on the outside temperature and on whether or not your home has insulating glass or storm windows. If you don't have insulating glass or storm windows, you need lower humidity to avoid condensation. You should also decrease the humidity when the outside air temperature goes down. As a general guideline, you might use the following humidities for different outside air temperatures: 35 percent for 20°F, 30 percent for 10°F, 25 percent for 0°F, 20 percent for —10°F, and 15 percent for —20°F. Of course, you can also adjust the humidity until you notice moisture condensing on the windows, then maintain it just below that level.

If you're in a climate where condensation is a concern, it's highly unlikely that your home is uninsulated. If it is, though, moisture may condense on the walls before it does on the windows, and this should tell you when to decrease the humidity. Excessive condensation should be avoided because of its deteriorating effect on window sills and walls.

For a small amount of money you can buy a humidity meter or hygrometer which will be very helpful in checking the humidity level of your home.

INSIDE RELATIVE HUMIDITY ABOVE WHICH WINDOW CONDENSATION OCCURS

| | Inside Relative Humidity | |
| | Single Pane Glass | Insulating Glass or Storm Window |
Outside Air		
+ 20°	29%	55%
+ 10°F	21%	44%
0°F	16%	43%
— 10°F	11%	39%
—20°F	8%	34%

Check the accuracy of the hygrometer by placing it in your bathroom while you're taking a shower. At the end of your shower, the hydrometer should read close to 100 percent.

Even without a hygrometer, you can check your house's humidity level by using a glass of ice water. With the house temperature at 72°–75°F, put three ice cubes in a glass of cold water. If moisture doesn't collect on the glass at the end of three minutes, you need additional humidification. Don't do this test in your kitchen or the bathroom, incidentally — there may be excess moisture in either room, and the test won't mean anything. Also, don't forget to turn the thermostat down again after the test.

ADJUSTING RADIATORS AND REGISTERS

The registers and radiators in your home should be properly balanced or regulated to provide you with an even distribution of heat: improper balancing causes some areas to be too warm and others too cold. Dampers on the registers and valves on the radiators can be adjusted to provide the proper balance. Also, during the day you might wish to partly close registers or shut off radiators in areas which are warmed by sunlight. It is also important that return air registers be unobstructed.

For your radiators to work at top efficiency, they should be thoroughly cleaned of all dust and debris. You can do this by vacuuming them occasionally. You may also want to vacuum air registers and the inside of the ducts for as far as you can reach with your household vacuum cleaner. And if you install aluminum foil reflectors behind radiators, you can reduce the amount of heat being radiated to outside walls. Since much of this energy is lost, foil reflectors save you heat and money.

UNUSED ROOMS

If you have rooms which you're not using, isolate and insulate them as much as possible from the rest of the house. If you have a hot air heating system, close all registers in the room. If you have a hot water or steam system, and if the radiators in the room can be isolated from the rest of the system, they should be turned off and then drained to prevent them from freezing. If you have electric heat, turn off the thermostat in each unused room. Close all shades and drapes, and if you're not going to use the room for some time, put insulation on the inside of the door and install weather stripping around it. After closing the door, roll up a rug and place it across the bottom of the door on the outside to seal off cold air leaks.

GARAGE

If there's one luxury we can forego, it's the heated garage. It's true, of course, that you want to keep your car engine warm for easy starts in the winter, but you can do this by keeping a light bulb burning under the hood. You might also use an electric heater in the oil dipstick or in the water cooling system. You can buy either kind of heater at an auto supply store, and either will keep your engine warm for considerably less than it costs you to heat your garage.

We'll assume your garage is unheated. If it's an attached garage, though, how do you keep it from taking heat out of your house and sending your heating bills upward? Well, you can begin by keeping the garage door closed except when you're starting your car in the garage. (When the motor is running, of course, you keep the garage door open to avoid asphyxiation.) You also want to apply weather stripping around the door leading from the garage into the house, and you might even install a storm door or put insulation on the garage side of this door. If your garage has windows, make sure that they fit tight, sealing the edges with tape if necessary. Finally, try to close your garage door immediately after you park the car in the garage and shut off the engine. You want the heat (but not the carbon monoxide) from your engine to stay in the garage for as long as possible.

USING THE FIREPLACE

Our forefathers used fireplaces to heat their homes, and even today there is a certain charm and a physical warmth we feel as we sit around a crackling fire in the fireplace. The better the draft or draw, the better the fire. This draft, however, also sucks warm air from our houses. While we bask in the warmth of the fire, the warm air from our houses is going up our chimneys. As a result, heat losses can increase by as much as 20 percent while our fireplaces are operating. This loss is reduced somewhat by the heat generated by a fireplace fire, but most of this heat is also lost up the chimney.

If you're serious about doing all you can to conserve energy, don't use your fireplace. If you insist on using it even though it wastes fuel and may actually make your house colder, then at least try to reduce the heat losses through your chimney. When the fireplace is operating, close doors to adjacent rooms to keep warm air from being pulled out of them and wasted. When the fire is almost out, close the damper half way to reduce air flow up the chimney, and be sure to close it entirely when the fire is completely out. Stirring the ashes and making sure they're cold before closing the damper will keep you from accidentally filling the house with smoke.

It's easy to forget to close the damper, but you can jog your memory by considering an open damper to be like an open window, except not as noticeable. What the damper does is let warm air escape from your house, causing cold air to seep in through cracks in the walls, doors and windows. This way you don't notice any blast of cold air coming down the chimney, but your house becomes colder and your heating bill goes up.

SPACE HEATERS

Space heaters, as the name implies, are designed to heat only a limited area of a house. They heat the air only in their immediate vicinity and depend upon air circulation to carry the heat to other areas. This causes the air to be quite warm near the heater but much cooler when you get several feet further away, and therefore heaters are not an effective way to heat a house or even a large part of a house. A central heating system of any kind is much more efficient in distributing heat evenly throughout your house.

There are some situations where use of space heaters is warranted. If your central heating system doesn't have enough capacity to heat your home, you could use space heaters for supplementary heating. Or if you live in an area where the winters are rather mild, you may save money by buying space heaters instead of installing an expensive and seldom-used central heating system. Again, you may want space heaters to heat infrequently used facilities such as vacation cabins located in cold climates.

If you do use space heaters, select a model that has a small blower, or use a fan to help circulate the hot air near the heater to other parts of the area you're heating. There are also several other factors you should consider when buying a heater. First of all choose a heater which has adequate capacity but is not oversized for your needs. In general, a smaller unit operating almost continuously is more efficient than a large one operating infrequently. Gas and oil fired units usually require a vent stack to carry off exhaust gases and are therefore immobile units; electric heaters are portable and can be moved easily to various locations where heat may be needed. Unfortunately, electric heaters also use about twice as much fuel as gas and oil models. Unless portability of the heater is fairly important, then, you should get a model that runs on gas or oil.

There are small gas and oil heaters which don't need venting and are therefore portable. You have to be very careful with these, however, since their exhaust goes into the room they're heating. If they're not operating properly, they can generate lethal amounts of carbon monoxide.

Some space heaters rely primarily upon heat radiation for their effect. Electric heaters with exposed radiating coils and gas heaters with radiating flames fall into this category. These models do heat the air to some degree, but their major effect is to heat nearby individuals by direct radiation. Other units such as oil-fired space heaters, wall-mounted gas units and portable electric baseboard convectors, rely upon natural convection or circulation of the heated air. You should buy radiative units if your main purpose is to provide warmth for a few individuals without necessarily heating all of a room. Convective units are for rooms or larger areas.

HEATING YOUR SWIMMING POOL

If you own a swimming pool at all, you're more fortunate than many people; if you can heat your pool, you're even more fortunate. Since heating a pool requires tremendous amounts of energy let's discuss efficient and low cost ways of doing so.

Several years ago, swimming pools were concentrated in the West and South where the climate was such that they could be used almost year round. Now, however, pools are located throughout the country. Many pool owners in the colder parts of the country use heaters to keep the water at a comfortable level and/or to extend their period of usefulness in the fall. This heating uses large sums of energy. Electricity is also needed to operate the circulating pumps for the filtering systems. In the summer these pumps may use 600–800 kilowatt-hours of electricity which is enough to operate your dishwasher, washing machine, and the color television for an entire year. This energy is almost insignificant, however, when compared to the energy needed to heat the pool.

The energy needed to heat your pool depends both on the temperature levels you maintain and on the local weather conditions. A daily energy requirement of 300,000–500,000 Btus is not at all unusual. During the four month heating season this amounts to about one-half the energy required to heat a home in the midwest during the entire heating season.

If you presently have a heated pool or are contemplating the installation of one, there are several things you can do to reduce your energy requirements for heating.

26

In the first place, the water temperature should be kept as close to the air temperature as possible when the air is cool, as this reduces heat loss from the pool. (When the air is warmer than the water, of course, it helps heat the water). The amount of energy that can be lost from a pool is tremendous. When the air temperature is $5°F$ below the water temperature, for instance, a 20 ft × 40 ft pool will lose about 4000 Btus of energy in an hour. If the temperature difference is increased to $10°$, the heat loss more than doubles to approximately 8500 Btus per hour, and a $15°$ temperature difference will cause 15,000 Btus per hour to be lost. Heat loss overnight may easily be a quarter of a million Btus.

Heat loss from the pool is the result of three different factors: contact between water and cooler air, evaporation (evaporating water carries heat with it) and radiation. Lowering the temperature of the pool reduces heat loss to the air. Evaporation and radiation losses can be reduced by covering your pool when it's not in use. A good cover will not only reduce heat loss, but also heat your pool somewhat in the daytime by trapping heat from sunshine. This "greenhouse effect" will work even when the air outside is cooler than your pool.

You can buy a standard pool cover or build one yourself from a thin sheet of polyethylene plastic and a few pieces of aluminum tubing for a support frame. Either way, it's well worth the investment. A man in Michigan who uses a cover maintains his pool at about $80°F$ to $90°F$ from the end of May until the middle of September without any additional heating. His neighbors, who use no pool cover, spend \$150–200 a summer to heat their pool without extending their swimming season beyond his. For the sake of your budget, if not for the sake of the energy crisis, you owe it to yourself to follow his example and not theirs.

CHECKLIST FOR HEATING THE HOUSE

Furnace (Hot Air Systems)

> Is the burner or flame adjusted correctly?
> Is the combustion chamber clean?
> Have you cleaned or replaced the filters?
> Does the electronic filter operate correctly (if present)?
> Is the humidifier operating correctly (if present)?
> Do the blower and motor need oiling?
> Is the drive belt on the blower tight?
> Does the blower need to be cleaned?
> Have your checked the air ducts for leaks?
> Have air leaks in the ducts been sealed?
> Are the supply and return registers unobstructed?
> Has the air distribution system been balanced for even heating?

Furnace (Hot Water or Steam Systems)

> Is the burner or flame adjusted correctly?

Is the combustion chamber clean?
Have you checked pipes and radiators for leaks?
Have leaks been repaired?
Is air flow around your radiators unobstructed?
Have the outsides of your radiators been cleaned?
Have you checked for trapped air in radiators?
Is the slope of return lines correct to avoid traps?
Is the system balanced for even heat distribution?

Insulation

Can you add additional insulation in the ceiling?
If you're installing insulation, is the vapor barrier placed correctly?
If you're building a new house, have you planned on adequate insulation?

Insulating Glass and Storm Windows

Have you installed storm windows?
If you don't have storm windows, can you economically justify them?
Are the air vents in your storm windows open?
Can you install plastic sheets over windows for added insulation?
When airing out your home, do you open many windows at once for a short
time during the warmer part of the day?
Is your window air conditioner removed, or at least insulated and covered?

Weather Stripping and Caulking

Have you caulked around door and window frames?
Have you checked to see that glass panes are tight in window frames?
Is your unused milk delivery box sealed and insulated?
Is the door or wall mail slot sealed, or a mail receiving box installed?
Did you install weather stripping around doors and windows?
Are the air vents in crawl spaces partially closed off?
Is the interior moulding around doors and windows tight?
Is the door or hatch to the attic insulated and sealed?

Drapes and Shades

Do you close drapes and shades at night and on very cold or cloudy days?
Do you open drapes to let sunlight in?
Have you selected the proper color drape?
Are thermal liners removed during the winter (if present)?
Are venetian blinds set at the proper angle to let sunlight in but still trap air?
Have you installed deflectors on registers and radiators where necessary, to keep
warm air from behind drapes?

Thermostat

Have you determined the lowest comfortable setting?

Do you lower the thermostat in the evening?

Do you lower the setting when the house is vacant for a considerable length of time?

Have you considered installing a dual setting, timer-actuated thermostat?

Are thermostats in adjacent rooms set within 5° of cne another?

Has the calibration of your thermostat been checked?

Is the thermostat located in the proper place?

Have you shielded the thermostat from sunlight and heated air?

Humidity

Do you have high enough humidity in your home?

Do you use a humidifier?

Is the humidifier operating properly?

Do you use pans of water near registers and radiators instead of a humidifier?

Can you use your clothes dryer exhaust for humidification?

Have you checked for condensation on your windows?

Do you have a humidity meter (hygrometer)?

Registers and Radiators

Is the air flow from the registers balanced?

Is water and steam flow to the radiators balanced?

Are registers and radiators unobstructed?

Has all air been bled from the radiators?

Have radiators been cleaned of dust?

Can you install aluminum foil reflectors behind radiators?

Unused Rooms

Is the heat to unused rooms turned off?

Are shades and drapes closed?

Is the back of the door insulated?

Have you sealed around the door edges?

Did you place a rug across the bottom of the door to seal off cold air?

Garage

Have you shut off heat to your garage?

If necessary, did you install a light bulb or some other heating device under your hood to heat your car engine?

Do you keep garage doors closed when the car engine is not operating?

Do you close the door immediately after parking the car, so you trap engine heat?

If the garage is attached to the house:

— have you insulated the door leading into the house?

— have you installed weather stripping around the door leading into the house?

— can you install a storm door?

Fireplace

Have you considered not using, or at least reducing the use of the fireplace?

Is the damper closed when the fireplace is not in use?

Are doors to adjacent rooms closed when the fireplace is being used?

Do you close the damper half-way when the fire is almost out?

Do you check to make sure the fire is completely out before closing the damper tightly?

Space Heaters

Are you using space heaters properly for small areas or for supplemental heating?

If you're using several space heaters, can you install a central heating system instead?

Have you selected a space heater of the proper size?

Is the space heater vented to the outside, if required?

Is the flame adjusted properly?

Are you using radiating units for personal comfort and convective units for room heating?

Swimming Pools

Do you keep the water temperature at the lowest comfortable level?

Is the pool covered when not in use?

Is the pool cover transparent so that sunlight can penetrate it and heat the water?

CHAPTER III

COOLING THE HOUSE IN THE SUMMER

Air conditioning accounts for approximately 3 percent of the total energy used in the United States and is the third largest residential use of energy. Furthermore, the energy used for air conditioning is increasing at a rate of 15 percent each year, making it the fastest growing area of energy use in the country. Already, the widespread use of air conditioners leads to summer brown-outs and black-outs in many cities, and the problem may get worse unless we all begin to cut down on the energy we use to cool our homes.

Just as the amount of energy we need for heating depends upon the heat loss of our homes, the amount of energy used for air conditioning depends upon our heat gain or heat load. Generally, those factors which reduce the heat loss of a house in winter also reduce the heat gain in the summer. Insulating the house, installing storm windows or insulating glass, and sealing off air leaks, then, are all ways to reduce your house's heat gain and save yourself money on air conditioning.

If you live in an area with mild winters, you may have skipped the chapter on heating your home. If so, we recommend that you go back to it and read all the sections except those dealing with the furnace and the heating system. The sections on insulation, storm windows, weather stripping and the like will help you not only to heat your home, but to cool it as well.

However, there are a few suggestions in the chapter on heating which do need modifying before they can be applied during the summer. For instance, to reduce heating bills you should leave drapes and shades open on sunny days; just the opposite is true, of course, when you're cooling your home. When you want to air out your home in the summer, you still should open many doors and windows for a short time rather than a few for a longer time, but you should do this during the cool part of the day rather than during the hot part. (For the same reason, if you're going to introduce outside air into your air conditioning system, do it during the cool part of the day.) In the summer, be sure all vents leading to crawl spaces and to your attic are open, not closed, to prevent moisture condensation. And if you've been venting your dryer exhaust into your house during the winter, you should vent it outside in the summer to get rid of unwanted heat and humidity.

Now, what else can you do to reduce the heat gain of your home? To begin with, don't operate many of your appliances any longer than you have to. Electric irons, toasters, griddles, lights and similar devices will all add a fair amount of heat to your home. Cooking, of course, generates a lot of heat and moisture, although you can get rid of much of it by running your kitchen exhaust fan while you're cooking.

Secondly, try to shade your home as much as possible. Tall trees, shrubs and even wall ivy will absorb the sun's energy and make your house cooler. Ventilated awnings, porches and carports are also a good idea. Incidentally, if you're putting in a stone or concrete patio, try not to put it on either the west or the east side of your house — it'll reflect sunlight into your windows in the morning or in the evening.

Insulation has already been pretty well covered in the chapter on heating, so we won't say much about it here. You might note, though, that the proper insulation may be able to save you 25 to 40 percent on the energy that your air conditioner uses. If you don't have an air conditioner, insulation will still make you cooler. A fan which blows air out of your attic in the summer will make your insulation most effective, since it will remove the hot air that does accumulate in the top of your house.

AIR CONDITIONERS

There are two types of air conditioning systems available: central air conditioning systems, and those using window units in different rooms. Units in both systems work in about the same way, but the central system uses a series of ducts to distribute cool air from a central unit to rooms throughout the home, while the window unit uses an internal blower and natural air circulation to distribute cool air over a much smaller area. Most of the suggestions in this chapter can be applied to either type of system, although some recommendations are only good for central air conditioners. It should be obvious which suggestions are of only limited applicability.

Before we consider ways to make your air conditioner more efficient let's first look at how it works. Basically, your air conditioner consists of two sets of coils, a compressor, and a working fluid or refrigerant. One set of coils is called the cooling coils, or the evaporator. Cold refrigerant circulates through the inside of the evaporator, while room air is circulated over the outside. Since the refrigerant in the evaporator is so cold, it absorbs heat from the warm room air and therefore cools the room air. It also causes moisture to condense on the coils, and this dehumidifies the air.

Now the refrigerant has done your air conditioning for you, but in the process it has become somewhat warmer. The refrigerant is now compressed, raising its temperature and pressure, and it enters the second set of coils or condenser. The higher temperature causes the refrigerant to reject heat to outside air surrounding the condenser. The high pressure refrigerant is expanded rapidly after leaving the condenser which makes it cold again; and, then, back to the evaporator it goes. This isn't all there is to air conditioning — but it's all we need to know to come up with some practical conclusions on how to save energy.

First, remember that your air conditioner can't be 100 percent efficient. Because the compressor heats the refrigerant in the act of compressing it, your air conditioner has to give off more heat from the condenser than it absorbs through the evaporator. If you run the unit with the condenser inside your house, then, it's net effect will be to heat the house instead of cooling it.

Another thing to remember about your condenser is that the outside air it comes in contact with should be as cool as possible. The cooler the outside air, the more heat it can absorb from the condenser, and the more heat the air conditioner can remove from your home. If at all possible, then, you should put your condenser where it's shaded from the sun. If it's located on the north side of your house, the building itself may do the job. Otherwise you might look to trees, shrubs, or even a decorative wall or screen to shade the condenser.

Whatever shade source you use, make sure that it doesn't block the free flow of air through the condenser. The more air flowing over the coils, the more heat the condenser can get rid of. Make sure, then, that air flow around the unit isn't obstructed by walls, vines, shrubs, trees or other objects. Also, and especially if you have a central air conditioning system with the condenser near the ground, check the air passages regularly to see that they are free of leaves and grass clippings. These frequently contribute to condenser blockage and lower cooling efficiencies.

The filters in your air conditioner should also be cleaned regularly to insure proper air flow. Check your owner's manual for correct cleaning procedures. Whatever you do, don't operate your system without a filter — this can clog your evaporator air passages with dust and seriously impair the evaporator's cooling and dehumidifying function. In extreme cases, when air flow is sufficiently reduced, the evaporator's cooling capacity can become too large for the small air flow, so that moisture not only condenses but also freezes in the unit. This eventually leads to total blockage of the evaporator, and is more likely to occur in central air conditioning systems than in window air conditioners. Both types of air conditioning systems are normally equipped with filters.

Keeping the humidity down is an important function of your air conditioner — in fact, it's more important to your comfort than cooling the air is. If you buy the wrong sized air conditioner, though, you won't get much humidity control, and your air cooling may suffer as well. If your air conditioner is too small, for instance, it will operate almost continuously without either removing enough moisture from the air or cooling it adequately. A unit that's too large, on the other hand, operates so infrequently that it cools the air but doesn't remove much moisture. In a unit which is really oversized, the cooling capacity of the evaporator may be so great that it freezes moisture from the air and blocks its own air flow. Again, this is more likely to occur in central air conditioning systems than in window units.

What's the right size air conditioner to buy for your home? Well, the capacity of an air conditioner is rated by the cooling capacity of its evaporator as expressed in Btus per hour. (Occasionally, you hear capacities expressed in tons. One ton of capacity is equal to 12,000 Btus per hour.) The air conditioner capacity that you need depends upon the heat gain in the portion of your house that will be serviced by the unit. Generally, an 18,000 Btu per hour unit is big enough to cool four rooms in an average insulated house. You might want to ask an air conditioning contractor for a more precise estimate of the capacity needed for your home.

Once you know what size unit you want, shop around to get the model with the highest efficiency. The efficiency or "energy effective ratio" of an air conditioner is measured by the ratio of Btus per hour it removes from your home versus the amount of the energy it consumes. Efficiencies vary: electric air conditioners have efficiencies ranging from 4.6 Btus per hour to 12.2 Btus per hour versus one watt of input. Gas-operated units are much less efficient. In spite of the low efficiency of power plants in producing electricity, electric air conditioners still use less energy than gas-air conditioners.

If your air conditioner is the right size, it won't run continuously. The blower may operate all the time, (depending upon your particular installation), but the compressor should only run intermittently. Your compressor may run continuously, however, on very hot days. If so, set the thermostat to a higher level, perhaps 80°F. The system will still remove moisture from the air, and this will keep you fairly comfortable even though the temperature is slightly higher. You should also consider setting your thermostat 5° to 10° higher whenever your house is going to be vacant for a period of time. Each degree you set the thermostat above 70° will save you about 4 percent in energy requirements. If you're going to be away from home for several days, of course, shut the air conditioning off entirely.

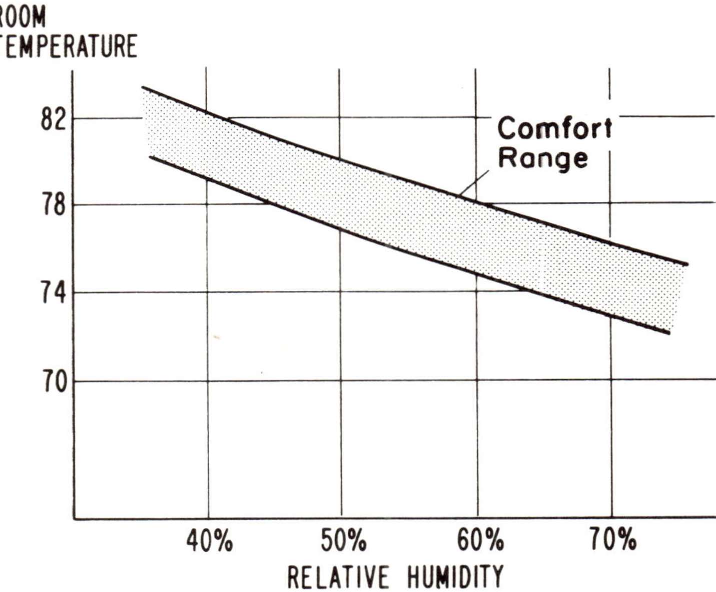

Recommended Room Temperatures and Humidity in the Summer.

If you use window air conditioning units, put thermometers in various parts of your house to check for overcooling. Overcooling wastes energy and money, and you should avoid it. Also, make sure your furnace thermostat is turned as low as possible whenever your window units are operating. More than one homeowner has awakened in the middle of the night to find his furnace and his air conditioners working against each other. If you have a central air conditioning system your furnace and air conditioner are probably connected, and you don't have to worry about this problem.

The ducts in central air conditioning systems should be checked periodically to see that all joints are airtight and no cool air is being lost. These ducts should also be insulated, both to prevent heating of the cooled air and to prevent moisture condensation on the ducts. The registers in the system should be unobstructed: if necessary, use deflectors to keep drapes from interfering with air flow. The air distribution system should also be checked for proper balance so that all areas are cooled evenly. Registers in the basement should be closed.

Routine maintenance will prolong the life of your air conditioner and make it more efficient. Evaporator and condenser coils should be cleaned annually, with the vacuum cleaner if possible. The grill work surrounding the condenser should be clean and unobstructed, and you may have to clean it fairly frequently if the air surrounding it is extremely dirty and dusty. Blower motors and fans should be oiled in accordance with the manufacturers' recommendations. Also, check to see that the blower fan belt in your central system is tight: it should move no more than half an inch when you press it with your finger. If your system seems to have less capacity than it used to, perhaps it has a leak and is losing refrigerant. Have an air conditioning contractor check for leaks and add refrigerant if necessary.

DEHUMIDIFIERS

A dehumidifier operates on the same basic principle as an air conditioner, but its use and installation are different. The dehumidifier is located entirely in your home, with no outside condenser, so it removes moisture from your air but can't cool the air at all. In fact, it heats the air slightly, and you should only use it when the air in your house is very humid but the temperature is moderate.

When you're using the dehumidifier, your home should be closed so moist air can't enter from the outside. You should locate the dehumidifier in or near areas of maximum moisture in your home, such as your kitchen, and you might want to use a fan or two to help mix and circulate the dehumidified air through other areas of the house. You can also use exhaust fans in the kitchen and bathroom to expel moist air from these areas.

The main purpose of a dehumidifier is not to provide for your personal comfort but rather to remove moisture from the air in damp areas such as your basement. This moisture can lead to deterioration of items stored in the basement. You should operate your basement dehumidifier even if you have air conditioning: this helps remove moisture which might otherwise infiltrate into the living area of your house and place an extra load on your air conditioning system.

A dehumidifier uses approximately 375 kwh of electricity per year. This is about the same amount used by your black and white television set or your furnace blower in a year.

FANS

If you use fans properly, you can provide a good deal of cooling for your home for a lot less money than you'd spend on an air conditioning system, and you can save energy as well. The one thing to remember, of course, is that you should use the fan to pull outside air into the house only when the outside air is cooler than the inside air.

During the daytime for instance, you neither want to blow outside air into your home nor to push inside air out the window. Air that goes out the window will just make other outside air flow into the house from other windows and doors. Shut the house tightly, then, draw the shades, and use the fan only to move air around inside the house.

At night, of course, you want to pull cool air into your home. There are two ways to do this. If you have the fan pull air in through a window in one room, it will cool that room first. Then, as cool air from that room spreads to other rooms, it will push the hot air in those rooms out the open windows. The rooms without a fan will thus cool down, though very slowly.

Suppose, however, that you have the fan blow hot air out of a window in one room, and leave windows in other rooms open. Cool air will be sucked in the windows of those rooms, cooling them first and then eventually reaching the room where the fan is. This is usually the best way to run the fan, since it pulls cool air into your house through several windows at once.

Several airflow and cooling patterns are possible in your home, depending on where you put the fan and which windows you open. For example you might set the fan to blow out a bedroom window and open windows only in your living room and kitchen. (The doors between these rooms, of course, must be open.) In this case cool air will be pulled through the living room and kitchen causing them to cool first; then as the cooler air reaches the bedroom this room will also be cooled. Keep in mind that the larger the space you're trying to cool, the longer it will take — and the area where the exhaust fan is located will be the last to cool. If you're using a window-mounted fan, you might want to install mounting brackets in more than one window. This way you can place the fan to blow out a bedroom window in order to cool the living room, then, when you go to bed, you can put the fan in the living room window in order to cool the bedroom. This may seem like doing things the hard way, but it cools more of your house, and it also keeps the fan in a room where its noise won't bother you. With a window-mounted fan, try to seal off the area between the window frame and outer case of the fan: this cuts down on the tendency for the air which has just been blown through the fan to be pulled around to the air inlet side again.

The size and the number of fans you need will depend on the size of your home and on how rapidly you want to cool it at night. Fan capacities are normally expressed as the amount of air they can circulate in cubic feet per minute. An average 20 inch fan will handle about 2400 cubic feet of air per minute. This is about one-fifth of the air volume of a house having 1500 square feet of living area, so a 20 inch fan would produce a complete air change in five minutes in such a house if air flowed evenly in all parts of the house. This isn't the case, unfortunately, in most houses. Within each room warmer air tends to collect near the ceiling, and cool air must be circulated for some time before it displaces the warmer air.

Fans really do save energy when compared to air conditioners. The average window fan, for instance, consumes approximately 200 watts, and a small circulating or oscillating fan uses about 90 watts. Running your window fan for five hours, then, requires one kwh of electricity — the energy you need to operate a 18,000 Btu per hour window air conditioner for about twenty minutes. And when they're properly used, fans can provide adequate comfort except in cases of very high humidity or temperature. So even if you have air conditioning, you might consider using a fan for a major portion of your cooling needs. It'll save you a lot of money and energy.

Wrong

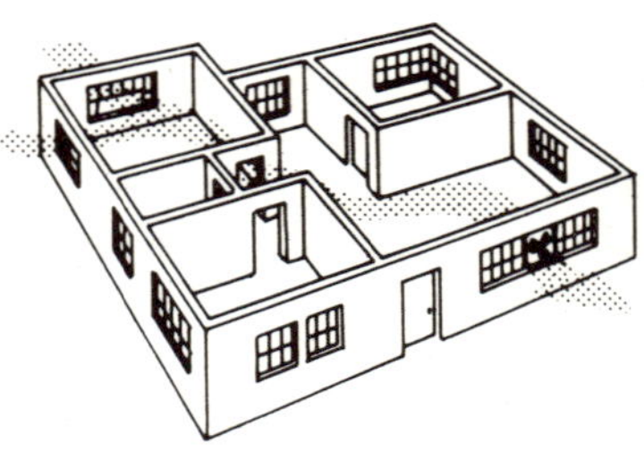

Right

Cooling a Room with a Fan.

DRAPES AND SHADES

When unshaded windows are exposed to sunlight, they can add significantly to the heat gain of your home. The amount of the sun's heat which enters through a glass window, for instance, is thirty-five times that which passes through a section of insulated wall of the same size. Insulated glass and storm windows resist the heat gain from warm air, but the sun's radiant heat passes through them easily, and the only real way to combat the sun's heat is to shade your windows.

One way you can do this is by installing ventilated awnings over the windows. Another and easier way is to use drapes and shades. As was pointed out in the chapter on heating, light colored drapes and curtains (and white shades) are best because they reflect the sun's radiant heat. Dark colored drapes and shades absorb this heat. You might wish to install a separate drapery liner designed specifically to reflect sunlight: these liners are available at most places where drapes are sold.

The chapter on heating indicated that the air trapped between the drape and window was most important. Not so for cooling the house. It's more important to reflect the sunlight back outside.

CHECKLIST FOR COOLING THE HOUSE

General

> Have you read the sections in the heating chapter dealing with insulation, insulating glass and storm windows, weather stripping and caulking, drapes and shades and unusued rooms?
> Do you leave storm windows on in summer?
> When airing out the house, do you open many windows at once during the cooler part of the day?
> Do you introduce outside air through your air conditioner during the cooler part of the day?
> Are vents in the crawl space and attic open?
> Is the dryer vented to the outside?
> Are you operating heat-generating appliances for as short a time as possible?
> Do you use the kitchen exhaust fan when cooking?
> Is your home shaded by trees and awnings?
> Have you avoided concrete patios and drives next to windows on the east and west sides of the house?
> Can you insall an attic ventilating fan?

Air Conditioners

> Is the condenser unobstructed so that air may circulate freely?
> Is the condenser shaded from direct sunlight?
> Are the filters cleaned?
> Have you selected a unit with proper capacity?
> Have you selected a unit that operates with maximum efficiency?
> On very hot days, do you set the thermostat higher?
> When the house is vacant for a considerable length of time, do you set the thermostat higher?
> Is your furnace turned off?
> Are air ducts tight and leaks sealed?
> Are supply ducts insulated?
> Are the registers unobstructed so air flows freely?
> Did you install deflectors to keep cold air from blowing between drapes and windows?

Is the air distribution system balanced for even cooling?
Are registers in the basement closed?
Is the condenser of your window unit located outside the house?
Have you cleaned the evaporator and condenser coils?
Do the blower and motor need oiling?
Is the fan belt tight?
Does the system need more refrigerant?

Dehumidifiers

Is the dehumidifier located near areas of high moisture?
Do you use kitchen and bathroom exhaust fans for moist air removal?
Are the dehumidifier grills clean for unrestricted air circulation?

Fans

Are you using fans properly (exhausting warm air and pulling in cool air)?
Have you selected the proper air flow pattern?
Have you sealed around the window exhaust fan?
Do you have the right number of fans? Are they the right sizes?
Can you use a fan instead of your air conditioning on some days?

Drapes and Shades

Do you close drapes in windows exposed to sunshine during the daytime?
Have you selected drapes of the proper color?
Are you using reflective drape liners?
Do you draw shades during the daytime to keep out sunshine?
Are venetian blinds set at the proper angle to keep sunlight out?

CHAPTER IV

HOW TO SAVE ON THE ELECTRIC AND GAS BILLS

You must have noticed, no doubt with dismay, how your electric and gas bills have increased over the years. You probably wonder whether to blame the utility companies for their high rates or your children for leaving on the lights in their rooms. If so, you can stop wondering. Of course, the cost of electricity and gas has gone up, like almost everything else. And lights left burning around the house do use up electricity. However, your new appliances are mostly responsible for the surge in your energy consumption. Thirty years ago, when electricity consumption was about one tenth what it is today, few families had air conditioning, television, dishwashers, dryers, freezers, and the many other appliances so central to the modern household. All these appliances are a considerable load on our energy sources and contribute to the present energy shortage.

An extremist might say that the only way to eliminate the problem is to eliminate your appliances. But such drastic solutions are not necessary. You can save a lot of energy-and money-by selecting your appliances with care and by using them efficiently. Good habits in operating your appliances help you save energy in more ways than one. They reduce your gas and electricity consumption and, at the same time, add to the life of the appliance. Since it takes energy to manufacture the appliance, you save additional energy by prolonging its life.

There are many ways in which you can keep down the operating cost of your household appliances without giving up much of your comfort and convenience. All appliances will function best if installed, maintained and operated according to the manufacturer's recommendations. Keep a reference file of all instructions that come with your new appliance. Be sure to follow the instructions when you install that appliance. It is also a good idea to check all the appliances which you have every so often, to see if they are adjusted properly. An incorrect temperature setting on the dishwasher may waste hot water and result in dirty dishes. An improper adjustment on the humidifier may cost you electricity and also make your windows steam up. A poor adjustment on any one appliance may not affect your bill by very much, but your total energy consumption is made up of many small parts. Every little bit of saving helps.

MAJOR APPLIANCES

After heating and cooling, the largest users of energy in your home are the water heater, washer, dryer, dishwasher, refrigerator, freezer, television and cooking range. Because their energy demand is high, these appliances may waste significant amounts of energy. But they also offer the most potential for savings. It is worth your while, therefore, to select major appliances carefully when you are buying them. If you've already invested in appliances that burn a lot of energy, you can still save by changing the way you operate them.

Water Heater

Your water heater requires little attention and provides hot water for many years. Nevertheless, while it stands unobtrusively in the basement, it consumes generous portions of energy. An average water heater uses about 4500 kwh a year, which may be as much as one half or one third of the total energy used in your home. Any saving which affects the water heater will make a significant contribution towards reducing your utility bill.

When you are in the market for a new water heater, first decide what size heater to buy and whether to get an electric or gas model. If natural gas is available to your house, a natural gas heater is preferable to an electric one. Natural gas water heaters have lower efficiencies than electric ones, but remember that energy is also used to produce electricity, and electric plants are far less efficient than your water heater. If you were to add up the energy losses in the plant and in your heater, you would find that the total energy requirement of an electric water heater is higher than that of a gas heater of comparable size.

The size of the heater has a significant effect on your energy consumption. The larger the heater, the more energy is needed to keep it running. Your local utility company can tell you which size heater will satisfy your family's requirements. The correct size heater will provide all the hot water you need, and it would be a mistake to buy a larger one. A bigger heater just burns more energy — and money — without any added benefit to you.

ELECTRIC CONSUMPTION OF MAJOR APPLIANCES

	Average Wattage	Estimated kwh Used Yearly
Clothes dryer	4850	1000
Dishwasher	1200	350
Freezer		
15 cu ft	340	1200
15 cu ft, frostless	440	1760
Radio	70	85
Radio-Phonograph	110	110
Refrigerator		
12 cu ft	240	730
12 cu ft, frostless	320	1230
14 cu ft	325	1130
14 cu ft, frostless	615	1830
17 cu ft	400	1250
17 cu ft frostless	700	2000
Television		
Color	330	500
B & W	235	360
Washing machine		
automatic	510	100
non-automatic	290	75
Water heater		
standard	2500	4200
quick recovery	4500	4800

The size of the heater you should get depends on the size of your family and on the number of bathrooms in the house. It also depends on how fast you use up your hot water. The tank inside the heater contains a certain amount of water. Once you have used up all the hot water you must wait until the cold water in the tank is heated up again. You may avoid running out of hot water either by having a larger water heater or by using a smaller heater with "quick recovery". Quick recovery provides an additional burst of energy as the hot water is being used up. This extra energy raises the water temperature quickly. Unfortunately, this feature may increase your energy usage by as much as 600 kwh a year. This is enough electricity to power your washing machine, dishwasher, and humidifier combined.

Therefore, you should try to get a standard model without quick recovery. Such a model should keep you well supplied with hot water at all times, unless everyone in the family decides to take a shower just as you are starting your fifth load of laundry.

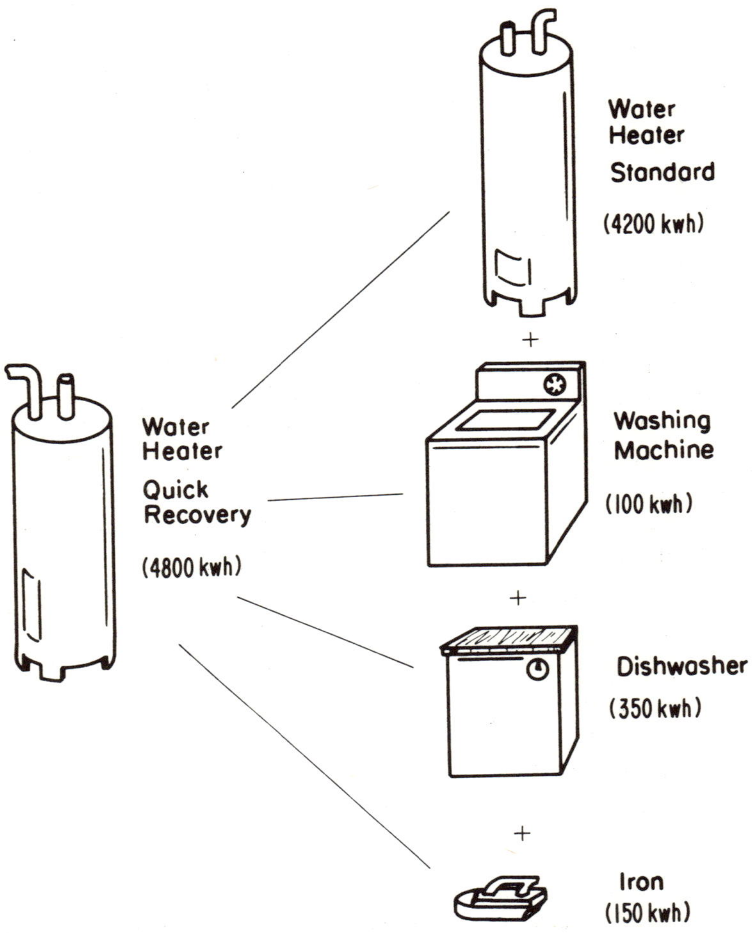

Energy use of a quick recovery water heater
(kwh represents approximate yearly consumption).

If your water heater had an efficiency of 100%, all the energy input to the heater would go toward raising the water temperature. In practice, some of this energy is lost to the air around the heater and to the air around the hot water pipes. Any reduction in this wasted heat is a direct saving of energy and money.

The amount of heat lost from the water heater depends on a number of factors: the size of the heater, the thickness of the insulation, and the temperature difference between the water and the air. A smaller water heater loses less heat than a larger one. This is one more good reason to buy as small a heater as you can get by on.

Once you have your heater installed you cannot do much about its size, unless you are willing to trade it in for a new one. But you can still reduce losses from your present heater by wrapping added insulation around the outside of the unit. A 1 inch thick layer of insulation will save you about 190 kwh per year. A four inch thick layer may save as much as 380 kwh. This is nearly 10 percent of the total energy used by your heater. Since the insulation is inexpensive you can recover its cost within about a year.

The temperature difference between the water and the air is also an important factor in the amount of energy that is wasted. The bigger this temperature difference, the larger the waste. In other words, for best economy keep the water temperature low and the room temperature around the heater high. We'll look at the question of water temperature shortly, but first let's examine the effect of the room temperature on energy loss. If the temperature of the room in which you keep your water heater is low, the losses from the heater are high. Install the heater, therefore, in a heated area. A water heater located in an unheated garage or in a cold basement increases your operating costs because of the large energy loss to the surroundings.

The water also cools down in the pipes. You notice this, for example, when you open a faucet which has not been turned on for awhile. The water coming out of the faucet is cold at first, and becomes gradually warmer as hot water fills up the pipes. The heat losses from the pipes are far from negligible and in fact can be quite sizeable. The heat loss from a copper pipe 10 feet long and 1 inch in diameter may be as much as 400 kwh per year. This is about 10 percent of your total water heating bill, or enough electricity to run your dishwasher for an entire year.

The heat loss from any pipe is directly proportional to the length of the pipe. Under the same conditions, a 10 feet long pipe loses about twice as much heat as a 5 feet long pipe, and a 20 feet long pipe loses twice as much heat as a 10 feet long one. Consequently, when installing a heater, you should reduce the length of the pipes by placing the water heater as near as you can to the outlet where most of the hot water is used.

If your heater is already installed you can still save a considerable amount of energy by insulating the pipes. Ready-made insulation is available in plumbing supply stores and can be installed simply. A 1 inch thick insulation cuts the losses for a 25 foot long pipe to less than 200 kwh per year. For a 1 inch diameter pipe this means a yearly saving of about 700 kwh. For smaller diameter pipes the saving is less but is still significant. For a 3/4 inch 25 foot long pipe the insulation reduces the losses from about 780 kwh per year to 200 kwh. For a 1/2 inch pipe insulation

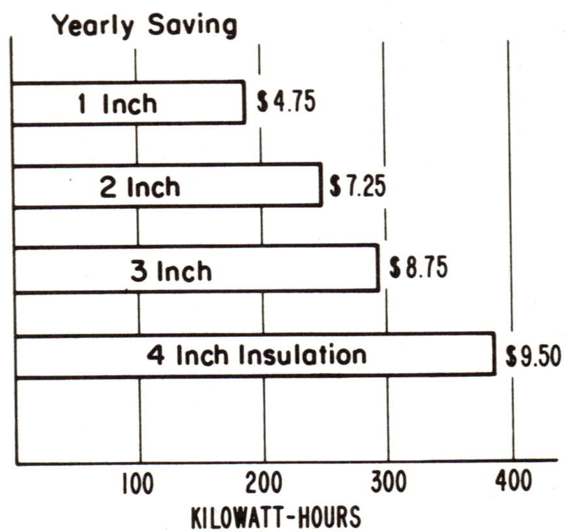

Savings if Insulation is Added Around the Water Heater.

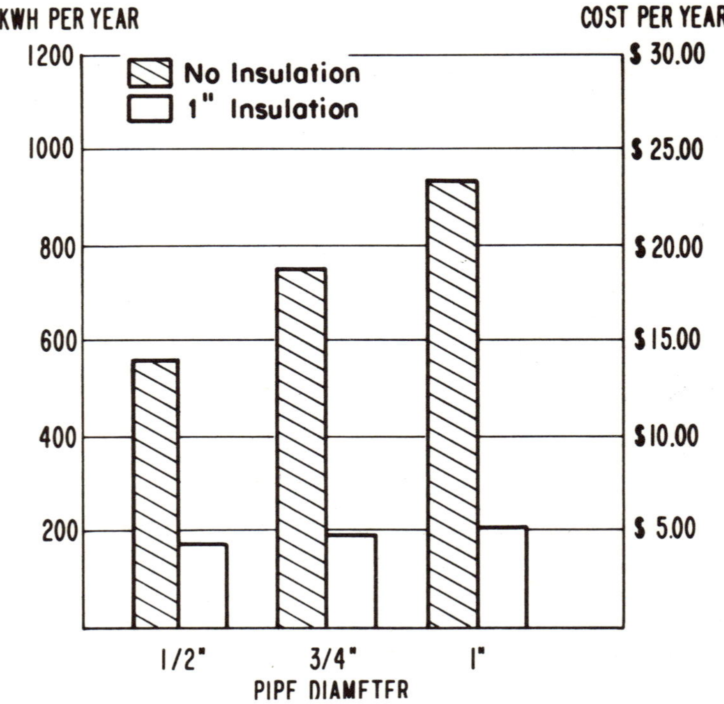

Heat Loss from a Hot Water Pipe (25 foot long, Copper).

reduces heat losses from about 580 kwh to 180 kwh per year. As you see, the diameter of the pipe is an important factor in the amount of heat lost. The bigger the pipe, the bigger the loss.

To minimize losses, then, keep the pipes short, small and insulated. This is also true for plastic pipes even though less heat is lost from these than from comparable copper pipes.

Now, remember what we said about water temperature? Well, your water heater has a built in thermostat which controls the water temperature. Be sure to set this thermostat at around 150°F. If the temperature is too low you will use too much hot water and your heater will be operating continuously to keep up with the demand. If you set it too high, then you are wasting too much energy through heat loss to the surrounding air. Since it takes energy to keep the water warm even when it is not used, turn down the thermostat twenty to thirty degrees when you are away from your home for a day or two. On longer trips turn the heater off completely. If you turn on the heater again when you arrive home, you will have hot water within a few minutes. Make sure that you follow the manufacturer's instructions whenever you adjust the controls on the water heater.

Although your heater needs very little maintenance, you should pay some attention to it. At least monthly drain off 1–2 gallons from the bottom. This helps keep rust and sediment out of the tank. For gas heaters also check the burner, and make sure it is adjusted properly.

Washing Machine

The clothes washer is one of your most important labor-saving appliances. Doing laundry without it is a real chore. Fortunately, automatic washers use only about 100 kwh a year, and with some care, you can reduce even this amount. The first rule for energy saving is never to use more hot water than necessary. You might even try a cold water detergent — in which case, of course, you do not need any hot water. In Japan, where energy has always been a scarce commodity, almost everyone launders with cold water.

The amount of hot water depends on your laundering technique. Obviously, the more you run the machine, the more hot water you use. So don't wash half loads. Save up enough laundry for a full load, while making sure not to overload the machine. An overloaded machine results in a poor wash, in wrinkled clothes and, possibly, in a blown fuse. If you are in a hurry wash small items, like stockings or socks, by hand.

Follow the washing instructions recommended by the manufacture. Sort out the clothes before laundering them and use the proper temperature for each load of wash. Use hot water for whites and for extra dirty clothes, warm water for colored materials and for lightly soiled garments.

Adjust your washing time for each load. A heavy load needs a longer washing time than a light load. Polyesters release dirt easily and can be washed quicker.

Don't be wasteful and do not wash longer than recommended. Contrary to what you might think, a longer wash cycle will not result in cleaner clothes. In fact, during the extra wash time dirt already removed from the clothes may find its way back into them again from the wash water.

When you are finished with the laundry close both the hot and cold water valves leading to the machine.

Clothes Dryer

A clothes dryer needs approximately 1000 kwh per year, or about ten times as much as a clothes washer. This is a fairly large portion of your overall energy consumption. The dryer, therefore, merits your special attention. By using it economically, you may save several hundred kwh yearly.

If natural gas is available to your house, buy a gas dryer. In general, they are more economical than electric ones. It takes 2 to 3 kwh of electricity to dry a load in an electric dryer. For the same load a gas dryer uses about 9 cubic feet of gas and 0.2 kwh of electricity. Operating costs for the two types of dryers are about the same, but it requires more energy to generate 2–3 kwh electricity than it takes to dry a load with a gas dryer.

Before buying a dryer size up your laundry needs. Estimate your load, and determine what kind of materials you dry most often. Pick a dryer which is the right size for your load. As with all appliances, larger units need more energy. If your laundry includes various types of fabric, buy a dryer with suitable temperature controls. These controls allow you to select the right temperature for each fabric. A higher than necessary temperature wastes energy, and may damage your clothes as well.

Since the dryer consumes a large amount of energy, use it sparingly. The most economical method, of course, is to dry the clothes on a line. Just two loads per week on the line saves you 20 kwh in one month. During the summer, when drying clothes outside is fairly easy, you can save up to 100–120 kwh, which is as much as the washer uses all year. In cold weather put up a line in the basement. And for smaller items you can use a portable drying rack.

Even apart from energy considerations, some items should always be dried on a line. Garments containing sponge rubber, foam rubber, fiberglass, and elastic should never go in the dryer. Bras and girdles may be dried in the dryer (at a low heat setting), but you might as well dry them on the line: it prolongs their life while saving energy. Do not put clothes which contain chemicals or wax in the dryer.

Always use your dryer on a full load but be careful not to overload the machine. Overloading increases drying time and wrinkles the clothes. Permanent press fabrics get wrinkled badly in an overloaded dryer. Remove these fabrics from the dryer as soon as the machine stops.

Different types of fabrics require different drying temperatures, so it's a good idea to separate the clothes according to their drying needs and dry similar types of fabrics together. As a rule, clothes that can be washed together can also be dried together. Experiment with your dryer to determine the best time and temperature settings for the various types of fabrics. Generally, heavy clothing needs the highest temperatures. Knits require lower temperature settings as indicated on some machines by "damp dry" or "less dry". Low heat should be used for lingerie. Once you've determined which setting to use for each fabric, take care not to overdry your laundry. Overdrying makes the clothing stiff and wastes energy.

For clothes with pockets turn the pockets inside out. This speeds up their drying. After each load clean out the lint trap. A dirty lint trap reduces the air flow and increases the drying time. The air flow is also hampered by a slipping belt on the blower. Check the belt periodically and tighten if necessary. Another area to inspect is the seal around the loading door. If the gasket is worn air is blown into the room instead of through the clothing.

Dishwasher

If you would be willing to give up your dishwasher you would save approximately 350 kwh of electricity a year. This may not be enough to make you return to the drudgery of washing dishes by hand, but it may convince you to operate your dishwasher economically.

For starters, always run the dishwasher on a full load. A partial load takes the same amount of hot water, electricity and soap as a full load. Also make sure that the water temperature is set correctly. For best results it should be between 140 and 160°F. Colder water does not get the dishes clean; hotter water bakes food particles onto your plates and cups. You can check the water temperature by placing a deep fat or candy thermometer inside the dishwasher and running it through a cycle. When you do this, remember to secure the thermometer so it won't rattle around and break. You don't want to be picking ground glass and mercury out of your food for the next month or two.

The water temperature will be too low in the dishwasher if you are using more hot water than your hot water heater can supply. Dishwashers with built in booster heaters help overcome this difficulty and raise the water temperature. Booster heaters also waste energy, though, and an adequate hot water supply is preferable. If your water supply cannot cope with all your appliances at once then turn on the dishwasher when it is not competing for hot water with the shower and the washing machine. Insulating the hot water pipes and adding additional insulation around the unit also helps.

Dishwashers are practically maintenance free. Do check the pipes and hoses for leaks, though, and see that the seal on the door is in good condition.

Refrigerators

Let's suppose, for the sake of argument, that you're buying a new refrigerator and want to get one that saves energy. How do you go about it? Well, before you start pricing different make and model refrigerators, first determine your family's requirements. Do you need a 17 cu ft frost-free refrigerator which gobbles up 2000 kwh per year, or could you do with a 14 cu ft model which needs only 1800 kwh? Even better, could your family get by with a 12 cu ft refrigerator which uses only 1200 kwh yearly? Also, is it worth the extra price for the frost-free model which consumes nearly 50 percent more electricity than an equivalent standard refrigerator? Undoubtedly, a standard unit needs to be defrosted periodically, but this is a small inconvenience considering that it saves you 500–700 kwh a year. This is enough electricity to run your color television, washing machine and radio for a whole year.

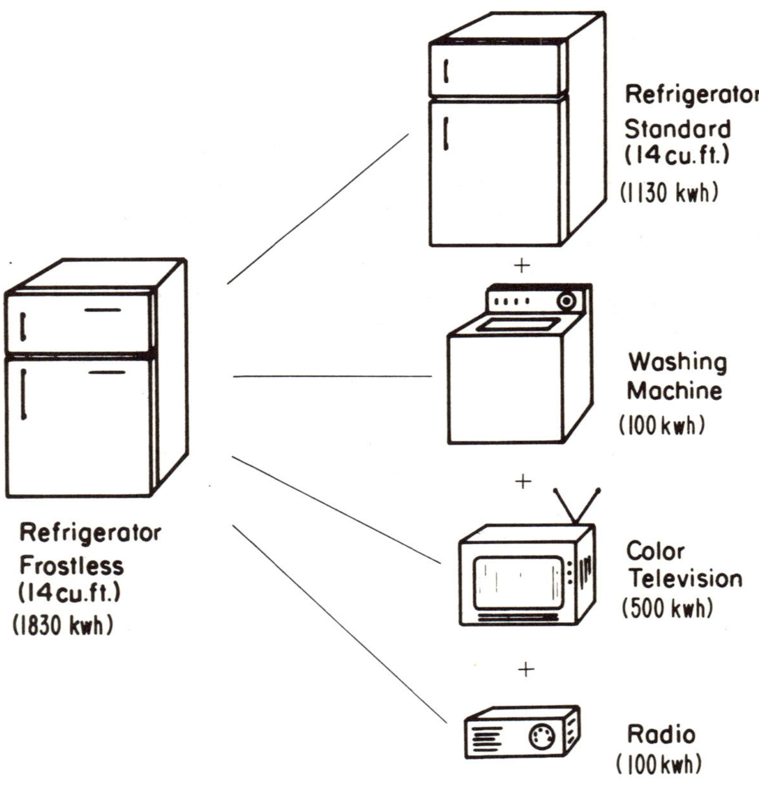

Electricity Used by a Frostless Refrigerator
(kwh represents approximate yearly consumption)

Frost-free units have an added disadvantage over standard models. In frostless refrigerators food dries out more quickly, and the food must be wrapped and covered more carefully.

Do you really need an automatic ice maker? This device uses electricity, takes up quite a bit of freezer space (1.5 cu ft) and, of course, needs a connection to the water supply. For your big parties you can get ice from the grocery store. For your everyday needs ice trays should provide an adequate supply.

When shopping for a refrigerator don't overlook the newer "efficiency" models, in which energy consumption is cut by improved design and insulation.

Your decision in buying a refrigerator will strongly be influenced by its cost. Interestingly enough, cheaper units are also more economical to run. The conventional single door models are the least expensive to buy and to operate. Top freezer models cost more and use more electricity. Refrigerators with bottom freezers are even more expensive and have higher electricity consumption. Side by side models cost the most in money and in energy.

Now let's suppose you've bought your new refrigerator — or that you're stuck with your old one. In either case, how do you operate it to save energy? To begin with, every refrigerator has an outlet, called condenser, where the heat taken out from the refrigerator is released. The condenser is located below or behind the refrigerator and is cooled by the surrounding air. The condenser must be cooled properly, for the refrigerator to function efficiently. If the condenser is overheated the cooling unit must work harder, resulting in higher energy consumption. Therefore, don't obstruct the condenser by cabinets that fit too closely to the refrigerator. If the condenser is on the back of the refrigerator, don't push the refrigerator too close to the wall. Keep the condenser clean by vacuuming it periodically to remove accumulated dust and dirt.

The warmer your kitchen the more energy is used by your refrigerator. Therefore, place the refrigerator in a cool place. Do not put it near the range, the radiator, hot air register, or in direct sunlight. If your kitchen arrangement is such that the refrigerator must stand opposite a sunny window draw the curtains and shades.

It is important to set the temperature inside the refrigerator at the proper level. Obviously, if the temperature is too high the food inside will spoil. At too low a setting the food freezes and energy is wasted. Most refrigerators have at least one temperature control, and many models have two: one for the refrigerator and one for the freezer compartment. If you have the choice of two controls, set the temperature in the refrigerator at 37°F and in the freezer at 0°F. If the freezer is fully packed lower the controls to ensure that the temperature is low enough even in those unavoidable warm spots. In units which have a single control for both the refrigerator and the freezer adjust the control so that the temperature in the fresh food space is about 37°F. The temperature in the center of the freezer should stay below 5°F. You can check the temperature with a refrigerator thermometer or with any ordinary room thermometer.

When you are away on a vacation turn the temperature control up. A "vacation setting" on the unit will do the same job. When you are not using the refrigerator for long periods of time, remove the perishables and unplug the unit. But be sure not to leave an unused refrigerator in an area where small children can get into it!

Every time you open the refrigerator door some cold air escapes. To minimize the loss of this air do not open the door unnecessarily and do not leave the door open too long. Decide what you need to take out or put in before you open the door. Knowing the location of the various items in the refrigerator helps you find what you want quickly and with little energy loss.

You can save energy by letting warm foods cool off before putting them in the refrigerator, but by doing so you run the risk of food poisoning. The bacteria which cause food to spoil grow rapidly at temperatures above 40°F. It is a good idea to cool foods as soon as possible, then, even though it uses extra energy. Whenever you are loading large amounts of unfrozen food into the freezer turn the temperature control to a lower setting. This will help to freeze the food quickly. Do not forget to turn the control back to the normal setting once all the food is frozen.

Wrap all food tightly in plastic wrappers, wax paper or foil. Cover all the jars and bottles containing liquids. In frost-free refrigerators it is particularly important that all the food is carefully wrapped and covered.

In a standard type refrigerator never let the ice build-up exceed 1/4 inch. A layer of ice on the cooling coils and cooling surfaces is an effective insulator which reduces the cooling power drastically.

For good economy the doors of the refrigerator should close airtight. To test the seal, close the door on a dollar bill. If the bill slides out when the door is shut, the door is loose and cold air is being wasted. Possible reasons for a loose door: the refrigerator is not level, the door hinges need tightening, or the gasket is worn. These problems are easy to fix. You can get replacement gaskets in the hardware store.

Freezer

Do you really need a freezer? Or could you just get along with the freezer compartment in your refrigerator? After all, separate freezers are expensive and they consume over 1000 kwh electricity yearly. Let's put this energy consumption in another light: a standard 15 cu ft freezer uses more electricity in a year than your radio, television, dishwasher, and washing machine combined. A 15 cu ft frostless freezer needs more electricity than a 14 cu ft standard refrigerator, plus a washing machine, plus a dishwasher.

Let's suppose you have a freezer, however. The general rules for economical operation of the freezer are the same as for a refrigerator. The unit should stand in a cool place, away from the furnace, water heater, radiator and sunshine. The condenser should be kept unobstructed and clean.

The ideal freezer temperature is 0°F, but it is difficult to maintain the entire freezer at this temperature. Some spots will be warmer, some possibly cooler. You can eliminate the warm spots by turning the temperature lower. You can also leave the temperature setting at 0°F, and place those items which you do not intend to keep long in the warmer locations.

Do not put too much food into the freezer all at once, because a large load raises the temperature. Within a 24 hour period you can freeze about 2 lb of food for each cubic feet of freezer space. This means you can load about 30 lb of food into a 15 cu ft freezer at one time.

Opening the door also raises the temperature, so open the door sparingly. Keep an inventory of the contents of the freezer and label the packages. This will save time in finding what you want.

Wrapping everything tightly in plastic wrappers, bags and alluminum foil helps prevent frost formation. Good wrapping is most important in frostless models.

The frost which forms on the cooling surfaces of the freezer must be removed. In frostless models this is done automatically at the cost of 400 to 500 kwh of electricity a year. You have to defrost a standard model yourself about twice a year. If you live in a humid climate you may need to defrost more frequently. In any case, don't let frost and ice build up in the freezer.

Check the seal on the door just as you do for a refrigerator. If the gasket is worn replace it.

Television and radio

During the past generation, television has replaced radio as the main entertainment in the home. And small wonder. It is much more enjoyable to watch a variety show, a ball game or a news program on television than it is to listen to it on the radio. It is also more costly, both in money and in energy. We all know about the difference in the purchase prices of a TV set and a radio, but few of us are aware of the big difference in energy consumption. A color television set uses about 500 kwh of electricity a year, while a radio requires only about 100 kwh. The 400 kwh difference is as much as a washing machine and a dishwasher use in about a year.

Nobody expects you to replace your color TV with a transistor radio because of the energy crisis. But if you're well enough off to be able to afford more than one television set, you might consider giving up that second or third set. And if you must have that extra TV, why not get a black and white one? It's cheaper, and it uses only half as much energy as a color set. Also, if you're buying a new TV, consider getting one with solid state circuitry. It will use 50 percent less energy than an old-fashioned set with tubes.

It goes without saying that the electricity consumption of a television set is directly proportional to the time it is turned on. A set which is on for 4 hours a day consumes about 500 kwh yearly regardless of whether you are watching it or not. If you turn the set off for two hours each day, you save 250 kwh in a year. One word of caution, though: don't turn the set on and off frequently — it damages the tube.

Range

The cooking range certainly falls into the category of major appliances, but we won't consider it here. We will come back to it later in the chapter when we look at the problems of efficient cooking.

PORTABLE APPLIANCES

It is one of the great wonders of our civilization how we succeeded in motorizing practically everything that moves. You do not have to use your own power anymore for such simple chores as brushing your teeth, shaving, or winding your clock. The electrical wizards put a motor even in your can opener and carving knife. This proliferation of various electrical gadgets certainly contributes to the energy shortage. But contrary to popular belief, they are not the source of all our energy problems. As a matter of fact, some of these appliances hardly make a dent in your electrical bill. Don't think, therefore, that you will conserve much energy by discarding your electric shaver or retiring your electric can opener. You would do better by giving up your electric blanket, humidifier, or attic fan.

If you're confused about which appliances to use and which to give up. look at the wattage of each one. The higher the wattage, the more electricity an appliance requires. But the wattage rating is not the whole story. An appliance uses electricity only while it is turned on. Therefore, a high wattage appliance used sparingly consumes less electricity than a low wattage appliance left on all the time. For example, an electric clock (with a rating of 2 watts) may consume more energy per year than a seldom-used floor polisher with a wattage rating 150 times as great. To determine total yearly energy consumption of an appliance, estimate the number of hours it is used in a year. Multiplying this estimate by the wattage and dividing the result by a thousand gives kwh of electricity used per year:

$$\text{kwh electricity per year} = \frac{(\text{wattage}) \times (\text{hours used per year})}{1000}$$

As a general rule, don't worry too much about appliances which use less than about 20 kwh yearly, but cut down on the use of those which are above 20 kwh. Among the higher users of electricity are the humidifier, dehumidifier. fan, and iron. The iron especially has a great appetite for energy, so don't use it more than you have to. If you are likely to forget to turn it off when you've finished ironing, put a timer on it.

Your electric blanket also uses a fair amount of electricity, but it takes less energy to keep the electric blanket going than it does to heat your bedroom. Turn down the thermostat, then, and turn on the blanket! Or, even better, turn down the thermostat, and replace the electric blanket with ordinary ones.

One nice thing about electrical appliances is that they are virtually maintenance-free. You have to oil a motor once in awhile and clean the appliance periodically,

PORTABLE APPLIANCES USING LESS THAN
20 KILOWATT-HOURS PER YEAR

	Average Wattage	Estimated kwh Used Yearly
Blender	390	15
Can opener	60	1
Clock	2	17
Defroster-refrigerator	350	4
Floor polisher	300	15
Hair clipper	35	1
Hair dryer	379	14
Heat lamp	250	13
Heating pad	65	10
Mixer	130	13
Sewing machine	75	11
Shaver	15	18
Sun lamp	280	16
Toothbrush	7	5
Vibrator	40	2

PORTABLE APPLIANCES USING OVER
20 KILOWATT-HOURS PER YEAR

	Average Wattage	Estimated kwh Used Yearly
Blanket	180	150
Dehumidifier	200	380
Fan		
circulating	90	40
window	200	170
Germicidal lamp	20	140
Heater (radiant)	1300	175
Humidifier	120	160
Iron	1100	145
Vacuum cleaner	630	45

but that's about all. Cleaning generally will not cut down on energy consumption of an appliance, with the exception of the vacuum cleaner. Its performance improves when you empty the dust bag.

LIGHTING

Compared to the rest of your electric appliances, electric lighting is a real bargain. It takes little energy to produce all the lighting your house needs. Nevertheless, it is important to use lighting correctly. The proper light in the proper place is best for your eyes and for your pocketbook. Here are some suggestions which can give you better illumination for less money:

First, select carefully the location where you intend to put a light. Decide what type of light you want and the wattage necessary to illuminate the area adequately. The "strength" of a light bulb is indicated by its wattage. Bulbs with higher wattage give more light. In selecting the bulb, bear in mind that different size light bulbs operate with different efficiences. Larger bulbs are generally more efficient than smaller ones. For example two 60 watt bulbs (for a total of 120 watts) provide the same light as a single 100 watt bulb. It is better, therefore, to use one large bulb in place of two smaller ones.

Fluorescent tubes are even more efficient than incandescent bulbs. A 40 watt fluorescent tube gives off the same amount of light as a 100 watt bulb, and lasts ten times longer. Since fluorescent tubes require less electricity than ordinary bulbs, they also give out less heat in hot weather. This means added comfort for you and a smaller load for your air conditioner.

When you are setting up a new lamp, experiment with different size bulbs to see which is the best. Needless to say, pick a bulb that is neither too bright nor too dim. Always shade the bulb. A naked lightbulb gives too much glare which is bad for your eyes. The shades should be made of light material to reflect the light, but the bulb should not be visible through the shade. The shade should be large enough to hide the bulb from you, regardless whether you are standing or sitting in the room. For floor or table lamps the bottom of the shade should be about 40–50 inches from the floor. Shades should also be open both at the top and the bottom, to allow the light to spread upwards as well as downwards.

Light a room evenly, avoiding a great contrast between light and dark areas. A bright spotlight in an otherwise dark room taxes your eyes. For this reason, turn on a light in the room in which you are watching television. Adjust the light so it does not reflect on the screen.

Do not leave your lights on unnecessarily. Lights left burning for long periods of time do add to your bill. For instance, a 100 watt bulb left on every night consumes 350 kwh a year. This is enough electricity to run your dishwasher for the entire year. Turn off the lights in unused rooms, except when you are away and your house is empty. It is wise then to leave some lights on to deter would-be prowlers. But don't leave the lights on all the time. This does nothing to increase your protection. Instead, lights left burning continuously in the same spot are a sign that no one is around. Time your lights, lighting different areas at different times to give the impression that your house is occupied. This goes for outdoor lighting too: turn all outdoor lights off at daybreak. If you are apt to forget, or are away from home often, install outdoor lights with built-in automatic timers. And lastly, use outdoor lighting only as needed. Use it for security, but not for ornamental purposes.

HOT AND COLD WATER

We are all accustomed to thinking of energy use in terms of electricity, gas, and oil, but we seldom think of water in the same way. Nevertheless, we are using energy when we use water, both hot or cold. It takes energy to pump the water, for one thing. If you have your own well this energy reflects in the electric consumption of your pump motor. If you live in a city it shows up on your water bill.

Reducing your water consumption saves energy. It also saves water which in many areas is in short supply, particularly during the summer months. The simplest way to save water is by using it moderately. Turn faucets all the way off after use. A running faucet obviously wastes water, but even a dripping faucet is a drain on your system. A faucet which drips one drop of water a second loses 200 gallons of water per month! You pay for this water, and if it happens to be the hot water faucet you also have to pay for heating all this wasted water. Fix leaky faucets.

There are other ways in which you can save water — for example, by not watering your lawn unnecessarily. According to gardening experts, it is better to water your lawn twice a week for 2 hours than every day for an hour, and this practice saves you 400—600 gallons a week. Also, a shower needs about 1/3 to 1/2 less water than a bath. If it is all the same to you, by all means take a shower. Finally, the water supply to your house is colder in the winter than in the summer, so it takes more energy to heat up the water in the winter. You can save energy by waiting until summer to do cleaning jobs which require lots of hot water.

HOW TO COOK EFFICIENTLY

Unless you are existing on cold cuts and on an occasional warmed-up can of soup, you are using a lot of energy in your kitchen. To be more specific, in preparing your food you use the equivalent of about 2000 kwh yearly (not counting the refrigerator and the freezer). What you would like to do, of course, is to reduce this without changing your eating habits or sacrificing the quality of your meals and dishes.

Saving in the kitchen begins at the range. Electric and gas ranges use different amounts of energy, with electric ranges generally requiring more. This may influence you when you are buying a new stove, but undoubtedly you aren't going to switch now just to save a few kilowatts. There are ways, though, in which you can economize with your present range.

One way to save energy is by using the right kinds of cooking utensils. For an electric range, utensils with flat bottoms are the best. The flat bottom is in contact with the heating element, so most of the heat from the element goes into your cooking. Pots and pans with curved bottoms or with rings or recesses allow heat to escape and waste energy. The most efficient utensils are those which have copper bottoms and stainless steel sides. The copper is a good conductor of heat and transmits it readily into the food. The stainless steel is a good insulator and helps keep the heat in the food.

Using the right sized utensil also saves energy. Heat is wasted if the heating element is larger than the pan, so match utensil size to heating element size. Use small pots and pans on small elements, larger ones on larger elements.

With gas ranges you do not have to be concerned so much about the shape of your utensils. Be sure, though, that the flame does not reach beyond the side of the pot or the pan. A flame which is larger than your utensil contributes little to your cooking, but a lot to your bill.

ELECTRIC CONSUMPTION OF COOKING APPLIANCES

	Average Wattage	Estimated kwh Used Yearly
Broiler	1450	100
Coffee maker	900	105
Cooker — eggs	500	14
Cooker — casserole	350	50
Cooker — pressure	1300	50
Deep fat fryer	1050	65
Frying pan	1200	186
Grill	1150	33
Hot plate	1250	90
Range	12200	1175
Roaster	1350	205
Tea kettle	1200	60
Toaster	1150	40
Toaster oven	1200	75
Waffle iron	1120	20

At the start of cooking extra energy is needed to heat up the utensil and the food. To supply this heat turn the heat to the highest position until the food begins to boil or sizzle. Then lower the heat. The heating element on an electric range retains heat for quite awhile, so turn the element off completely a few minutes before the dish is finished. Cast iron pots and pans also stay hot after the heat is turned off.

Defrost thoroughly all frozen foods before cooking, and during cooking cover all your pots and pans. This keeps the heat in the food, so your dish is ready more quickly and with less wasted energy.

The oven is a big help in the kitchen, but it is an energy gobbler. It pays, therefore, to use it wisely. Preheat the oven, but once the required temperature is reached put the food in right away. It is also a good idea to check the oven temperature with an oven thermometer. The built-in one may not be accurate. Also remember that you do not have to preheat the oven for roasting or broiling.

While it takes much energy to heat up an oven, little heat is needed to keep it hot. Thus, once you have fired up the oven use it to its full extent. Plant to prepare an entire meal in the oven. If you have extra space use it to bake cookies for a later occasion.

For best results do not open the door frequently. On the one hand heat escapes through the open door. On the other hand cold air gets in, which may raise havoc with your baking.

The oven works most efficiently with the proper utensils. Bright shiny utensils are better for baking than dark colored dishes. Aluminum is probably the best, but stainless steel, glass or ceramic utensils are also good. For pies glass plates are recommended.

For maximum efficiency the air should circulate freely between the dishes in an oven. Arrange the dishes so that they do not touch each other on the side of the oven. Stagger the dishes if you have them on more than one shelf.

Eventually every oven gets dirty. The temptation is to let the automatic oven cleaner do the cleaning for you, but each time you do this you use up about 3 kwh of energy. If you are really concerned about the energy problem, you should clean the oven by hand. Incidentally, the cost of cleaning the oven automatically and the cost of cleaning it with a chemical oven cleaner are nearly the same.

Use your oven only for what it is designed for: cooking. Ovens and ranges are very inefficient space heaters, so don't use them for warming up your kitchen. If you're cold, set up a small portable space heater or install a permanent radiator.

You might also wonder whether it is better to cook everything on the range, or whether you should utilize some of the single purpose appliances. The answer is: whenever you can use your electric griddle, frying pan, toaster, tea kettle, or coffee pot. These appliances use a lot of electricity. However, since they are designed for specific functions, they are more efficient than gas or electric stoves. For example, a range oven takes twice as much heat as a broiler. An electric tea kettle boils water for about half the electricity a range uses.

Microwave ovens are particularly good in saving time and energy. The manner in which microwave ovens cook is completely different from the traditional way of cooking. On a regular range, the heat is first transmitted to the dish; then it is transferred from the dish to the outside of the food; and, finally, it slowly penetrates the inside of the food. In microwave ovens heat waves are transmitted from one end of the oven to the other. These waves travel through the food, heating it up inside and outside all at once. The waves are designed to interact with the food but not with the dish, so energy is not wasted in heating up the dish. Because the cooking time is very short (seconds or minutes) and because little heat is wasted, microwave ovens do not heat up your kitchen. On a hot summer day this is a welcome advantage.

POWER TOOLS

If you live in a house you surely have a few small power tools to repair some of the unavoidable damage which occurs from time to time. If you like to build things you are also likely to have a few larger power tools, such as a power saw or a small lathe. All of these tools use electricity, but not too much if you use them moderately. The electricity consumption depends on the size of the motor and on the load. The consumption of a 1/4 horsepower drill, for instance, does not amount to much. A 1/2 hp saw used for a few minutes at a time does not require appreciable electricity either. The electric consumption of tools become significant only if they are operated for long periods of time. Even a small 1/2 hp motor needs 10 kwh in a week if it is running for 5 hours a day. As a rule do not use tools unnecessarily and turn them off when not in use.

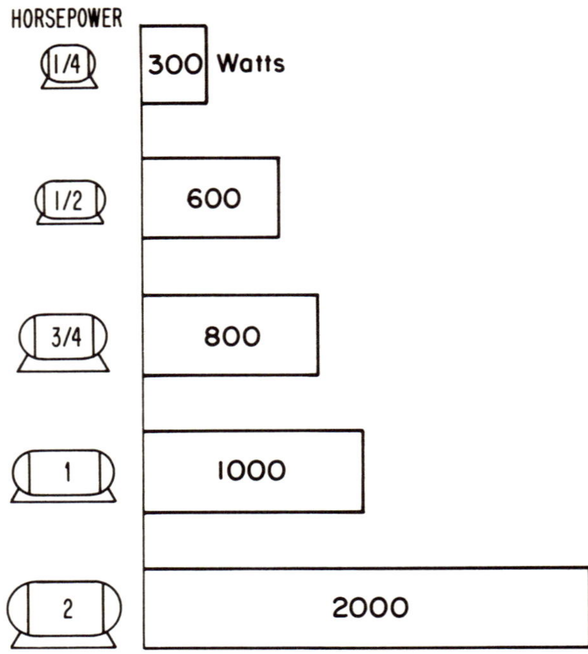

Approximate Electric Consumption of Small Electric Motors.

GARDEN EQUIPMENT

Before the advent of power mowers and hedge trimmers people seemed to be quite content in mowing their grass with push mowers and trimming their hedges with shears. Now on Sunday afternoons you can hear the buzzing of motorized mowers and trimmers all over your neighborhood. You might be bothered by the noise generated by these devices, but you need not be too concerned with their power consumption. A power mower will operate for 4–5 hours on a gallon of gasoline, which is enough to last through nearly half the summer. It takes even less energy to keep the bushes trimmed. Nevertheless, you do save energy by using your own power to cut the grass and trim the hedges. You might even enjoy the exercise. In the winter, when there's no grass to worry about, you can get the same exercise by shoveling snow and saving the gasoline that a snow-thrower would use. The amount of fuel saved isn't much — only about 2 or 3 gallons each winter — but if one million families would give up their lawn mowers and snow-throwers they would save 5 million gallons of gasoline a year. This is enough to keep 5000 cars on the road for a year!

WHEN TO USE ELECTRICAL APPLIANCES

The demand for electricity varies throughout the day. During "peak demand periods" the demand is high, while at other times it is relatively low. During the peak times, then, the electric company's generators may be straining to provide all the power that is needed, and during certain peak periods the generators even fall behind, causing brown-outs or black-outs. You can help relieve the strain on the generators by using your appliances in the off-peak periods, such as at night after 9 PM or in the morning before 9 AM. This is when you can do your dishes, laundry, heavy ironing, and baking without putting stress on the system. On hot summer days, help reduce the load on the generators by turning off the air conditioner for a couple of hours in the afternoon.

CHECKLIST FOR THE APPLIANCES

Water heater
    Is it the correct size?
    Is it in the proper location?
    Is it insulated adequately?
    Are the hot water pipes insulated?
    Is the thermostat set correctly?
    Have your drained off some water from the bottom?
    Have your checked the burner (gas heaters only)?

Washing machines

    Are you washing a full load?
    Are you overloading the tub?

Have you set the wash time correctly?
Are you using the correct water temperatures?
Have you turned off the valves after use?

Clothes dryer

Is it the correct size?
Does it have the proper controls?
Are you drying full loads only?
Is the dryer overloaded?
Are you drying similar fabrics together?
Are you using the correct drying temperatures and drying times?
Have you cleaned the lint trap?
Have you checked the belt and the door seals?
And finally, could you use a clothesline instead?

Dishwasher

Are you running it on a full load?
Is the water temperature set correctly?
Are you using it at the same time as the washing machine? Don't.
Are there any leaks in the pipes?
Is the door seal in good condition?

Refrigerators

Is it the right size and model for your needs?
Is it in the correct place?
Is the condenser unobstructed?
Have you cleaned the condenser?
Is the temperature set properly?
Are you opening the door unnecessarily?
Is the food wrapped and liquids covered?
Does it need defrosting?
Does the door shut tightly?

Freezer

Is it the right size for you?
Is it in a cool place?
Is the thermostat adjusted correctly?
Are you overloading the freezer?
Is the food tightly wrapped?
Are you opening the door too frequently?
Does it need defrosting?
Does the door close tightly?

Television

Do you need to use more than one set?
Can you replace a color set with a black and white one?
Are you turning the set off when it is not in use?

Portable Appliances

Have you checked the yearly electricity use of each of your appliances?
Are you using your appliances too much?
Have you turned your iron off?
Have you emptied the vacuum cleaner dust bag?

Lighting

Have you turned your lights off?
Are your lights in the right place?
Are the bulbs the correct size?
Do you use enough fluorescent tubes?
Are your lights shaded?
Are the rooms evenly lighted?
Are you using timers when you are away from home?
Are you using outdoor lights as ornaments?

Hot and Cold Water

Are you using water unnecessarily?
Have you fixed the leaky faucets?
Are you using a shower instead of a bath?

Cooking

Are you using the right shape, size and material utensils?
Are you matching the utensil to the burner?
Do you turn down the heat when the food is boiling or sizzling?
Do you cover pots and pans during cooking?
Do you defrost the food thoroughly before cooking?
Are you preheating the oven too long?
Is the oven at the right temperature?
Are you using the oven to its full extent?
Are you using the correct utensils in the oven?
Are the utensils arranged properly in the oven?
Have you stopped using the oven self cleaner?
Are you using the range to heat the kitchen?
Are you using your single purpose cooking appliances instead of the range?

Power Tools

Are you using them unnecessarily?
Have you turned them off after use?

Garden Equipment

Can you replace them by your own power?

General

Are you using your major electrical appliances in off-peak periods?

CHAPTER V

TIPS FOR THE MOBILE HOME OWNER

Mobile home owners have essentially the same opportunities as other home-owners to reduce heating and cooling costs. If you live in a mobile home, then, you should read the chapters on heating and cooling the home. Most of the information in those two chapters will also apply to your mobile home, although differences in construction may keep you from following certain suggestions.

For example, you won't be able to add additional insulation in your ceiling, nor are you likely to have access to heating ducts to check for air leaks. Well-constructed mobile homes, however, will have as much insulation as possible in the walls and ceilings, and the heating ducts will be sealed at the joints. The thickness of insulation possible in your mobile home is less than in a house, because your wall and ceiling spaces are narrower.

65

In addition to following the suggestions on how to heat and cool the home efficiently, you can take some additional steps to save energy in your mobile home. First, install skirting around the bottom of your trailer to prevent cold air from blowing under your floor and adding to your heat loss. You can add insulation to the back of the skirting for extra protection from heat loss, but, it's most important just to have the skirting. Uninsulated skirting made of aluminum siding or wood panels is fine. The skirting panels should fit as tightly as possible to restrict air movement, but remember also to provide at least two screened ventilators in the skirting: this prevents excessive moisture condensation under your mobile home. Another tip that'll save you money on heating: many mobile homes come equipped with ceiling ventilators. Be sure that these are closed tightly during the heating season. It also helps if you seal the edges of the ventilators with an insulating and sealing tape.

Now let's see how to cool your mobile home. The heat gain of a mobile home in the summer is strongly dependent on the color of the trailer and the amount of shading you get from the sun. A dark-colored trailer absorbs the sun's radiant heat and increases your heat gain, while a light-colored trailer reflects the sun's energy and has a lower heat gain. Shading also reduces the heat you gain from the sun. If possible, then, you should park your mobile home in an area shaded by trees. If your mobile home is permanently situated, you can plant climbing vines on trellises to provide shade, and of course you should install an awning. The awning should be set up so it shields your home against the sun during the middle of the day and the late afternoon.

Incidentally, shading is much more important for your mobile home than it is for a house. The house has thicker and heavier walls, and these walls act as a sort of heat sponge which soaks up heat during the day and then releases it slowly at night. Since the walls absorb heat, much of it is kept out of the house in the daytime. Your trailer, however, has thin walls which don't hold much heat. They cool off quickly at night, but they'll let you bake in the afternoon unless you make sure they're shaded.

How important are the effects of color and shade? Well, let's compare the increase in heat gain for various combinations. We'll compare against the best arrangement: a light-colored, completely shaded and well-insulated mobile home. If your unit is light colored but only has an awning for shade the heat gain will increase by 24 percent; without the awning it'll increase by 38 percent. If your mobile home is dark-colored and has no shade the heat gain will increase by 52 percent. Obviously, shade and color are important.

CHECKLIST FOR MOBILE HOME OWNERS

Heating and Air Conditioning

Have you read the chapters on heating and cooling the house?
Are windows and doors tight and weather stripped?

Do you open and close drapes and shades at proper times?
Are you setting the thermostat properly?
Are you using air conditioners and fans correctly?
Is your furnace adjusted and operating correctly?
Have you installed skirting around the bottom of the trailer?
Is your mobile home shaded adequately?
Are ceiling vents properly sealed in winter?

Appliances

Are you using the clothes dryer and washer properly?
Is the dishwasher being used correctly?
Are you using the refrigerator and freezer as recommended?
Are you using your appliances efficiently?
Can you use electric appliances in off-peak periods?

Others

Do you turn off unnecessary lights?
Is your lighting used efficiently?
Do you observe the rules of energy saving in cooking?
Are you wasting hot water?

A NOTE TO APARTMENT DWELLERS

Even if you live in an apartment you cannot avoid the discomforts of the energy crisis. The increased price of heating fuel is reflected in higher rent, and the brown-outs and black-outs imposed by the electric company affect your appliances, air conditioner and television. You, too, then, are ready to conserve energy. But how can you save energy when most of the obvious energy-saving measures are out of your control? Heating and hot water are provided by the landlord, and you don't feel like walking up to the 17th floor to cut down on the use of the elevator.

Admittedly, there's not much you can do to persuade a reluctant landlord to clean out the furnace, insulate the pipes, or put up storm windows. And walking upstairs is neither the most efficient nor the most convenient way of conserving electricity. Still, there are many ways in which you also can save energy. As a matter of fact, you can adopt most of the energy conservation methods used by people who own their own homes.

You can put up drapes to keep heat inside in the winter and outside in the summer. You can turn down the thermostat and shut off some radiators in cold weather. You can save electricity by raising the temperature control on your air conditioner or better still, by not turning on the air conditioner at all until the heat becomes bothersome. You can turn off unnecessary lights to reduce your electricity use, and further conserve electricity by using your appliances, refrigerator, and television efficiently. You can save energy by using hot and cold water sparingly and by employing the correct methods in doing your laundry. And you can save if you follow the rules of efficient cooking.

Details of each of these energy-saving steps are given in previous chapters. You may have already read these chapters and become familiar with them: if not, we recommend that you return to them, for they apply to you as well as to the homeowner. You will find many ideas in these chapters which will help you save energy without requiring that you tear up your apartment or otherwise bring your landlord's wrath upon you.

CHECKLIST FOR THE APARTMENT DWELLER

Heating and air conditioning

Are your windows and doors shut tightly?
Do you have weather stripping on your windows and doors?
Did you take out the air conditioner from the window in the winter?
Do you have drapes and shades on the windows?
Are you opening and closing the drapes and shades at the proper time?
Have your lowered the thermostat in the winter?
Have you shut off unnecessary radiators?
Have your raised the temperature control in the summer?
Are you using air conditioning unnecessarily?
Are you ventilating your apartment correctly?

Appliances

Are you using the clothes washer and dryer properly?
Are you using the dishwasher correctly?
Are you using the refrigerator and the freezer as recommended?
Are you using your portable appliances and television efficiently?
Are you using your electric appliances in off-peak periods?

Miscellaneous

Is your lighting arranged efficiently?
Have you turned off the unnecessary lights?
Do you observe the rules of energy saving in cooking?
Are you wasting hot or cold water?
Are you riding the elevator unnecessarily?

CHAPTER VII

SAVING ENERGY AT WORK

For most Americans, saving energy will begin at home — but there's no reason why it has to stop there. About half the nation's energy is used in places where people work, and according to some estimates 20 percent of this energy could be saved without any noticeable effect on productivity.

It's true that some of this savings could only be achieved by complicated and expensive changes in construction techniques, production methods, and the like. But a lot of the lost energy could be recovered by the same simple techniques which save energy in the home. In fact, many energy-conserving methods work better in offices and factories where larger amounts of energy are used.

For instance, unnecessary lights burning in your home consume a small, though noticeable, amount of electricity. Lights left on all night in an entire office building are real energy guzzlers. Brighter than necessary lighting of cafeterias, corridors and parking lots also consumes a fair amount of power — and therefore offers the possibility of a fair amount of energy conservation.

If you're in a position of authority in a factory or office building, you can make sure that some of this wasted energy is saved. You can also check to make sure that the building is insulated properly, that the doors are shut, that the windows are shaded, and that hot water pipes are covered to avoid heat loss. All of these steps should save the company money in the long run, although some may require a fair-sized initial investment.

Something you should avoid is trying to save energy by making your employees miserable. Turning down the thermostat a few degrees is a good idea — but don't turn it down to the point where the secretaries are freezing. Dimming lights in the corridors is another good idea — but not if your halls are so dark that they become safety hazards. Presumably you know by now that you won't get much work out of alienated, resentful employees, so be careful how much of the burden of energy conservation you ask them to bear.

If you're an employee, you don't have as much control over energy use as your boss does, but there are still things you can do to conserve power. Walking up or down a few flights of stairs instead of riding the elevator all the time, is one idea, though of course you aren't going to abandon the elevator for longer trips. Simply waiting a few minutes for other people to ride the elevator with you, rather than riding it all alone, is another way to save trips and save energy.

If you can, also try to turn off unused machinery at work. Practically all the machines in your office or factory use a lot of energy, and letting them run all day without using them is a waste. This doesn't mean you should keep turning them off and on again, though. Most machines take a fair amount of time and energy to warm up, so it's a good idea to wait until you have a lot of work to do on one machine, and then to do it all at once.

One office machine that most people find particularly important is the coffee percolator. There are times, after all, when it's all you have to get you through the day, and we're not suggesting that you do without it. But unfortunately, while you're gulping its coffee, the percolator is gulping energy — about 1 kw of electricity per hour. You can cut down on this energy by turning off the percolator when it's not needed. Instead of unplugging the percolator on your way out of the office at night, try turning it off an hour or two earlier. It'll still keep your coffee warm enough to get you through the rest of the day.

There are probably many other energy-saving methods which are particularly appropriate for the place where you work. Some of these may be simple: reducing the hot water temperature in the building, for instance, or dropping the pressure in the compressed air system. Other steps may require that the company invest some money in cleaning and tuning the boilers, insulating pipes and ducts, or converting furnaces to burn waste material. Just what methods are most economical for your plant will depend on local conditions, and here is where your knowledge of the plant and your imagination are superior to any advice we could give you. Look around — you may come up with energy-saving methods no one else has thought of.

CHECKLIST FOR SAVING AT WORK

Heating and Air Conditioning

> Are the rules for efficient heating and air conditioning observed?
> Are windows and doors shut tightly?
> Are the windows shaded properly?
> Are the thermostats set correctly both in the summer and in the winter?
> Is the heat shut off in unused rooms and buildings?
> Is the temperature reduced in non-work areas?
> Are buildings ventilated properly?

Electricity

> Is the lighting arranged efficiently?
> Are the lights dimmed in non-working areas?
> Are unnecessary lights turned off?
> Is the machinery in the workshops and in the offices used effectively?
> Are the machines turned off when not in use?
> Are unnecessary escalators and elevators shut down?
> Are elevators used efficiently?
> Is the coffee percolator running constantly?

Others

> Is the hot water supply at the right temperature?
> Is the compressed air supply at the correct pressure?
> Are hot water and compressed air used sparingly?
> Are pipes and ducts insulated?
> Are boilers, furnaces properly maintained?
> Could the production methods be improved to save energy?

And, last but not least,
> Have all possible energy saving steps been explored?

CHAPTER VIII

HOW TO GO MORE FOR LESS: SAVING GASOLINE

The automobile has become a part of our lives. It has offered us a freedom never known before, the freedom of movement. We have enjoyed this freedom tremendously and taken every advantage of it, and never for a moment did we think that the unlimited use of our cars would one day come to an end. Then the gasoline shortage hit us. It was like being rudely awakened from a dream. One day we had all the gasoline we wanted — cheaply. The next day we had to stand in line at our favorite service station to buy a few gallons at a high price.

It's nothing to worry about, you hope. Surely, the shortage is just about gone and you are able to travel about as freely as ever before. Well, the shortage may have indeed eased, but if we can trust our government's predictions it will be a long time before gasoline will be in abundant supply. And don't bother to look for that 30.9¢ per gallon sign again. Gasoline will certainly be no exception to the law that "the price that goes up mustn't come down again."

But the shortage and high price of gasoline are not the only reasons why we should curtail our seemingly irrepressible urge to drive. Our atmosphere is also suffering from the exhaust of the millions of automobiles on our roads. Emission control methods proposed for automobiles will ease the air pollution problem somewhat, but they do not present a complete solution. We will have to take further steps to clean up the air around us.

There are many different ways in which we can both help the gasoline shortage and minimize air pollution. The simplest method is also the most effective: stop driving. A car standing in the garage neither pollutes nor uses gasoline. It is worthwhile, therefore, to consider the alternatives to driving: walking or riding a bicycle. In some instances walking or bicycling not only saves gasoline and money but also saves time. In peak traffic hours, for instance it is often quicker to walk or pedal than to drive. In a recent race held in downtown Manhattan, both bicyclists and predestrians beat the automobile — and parking was not even part of the race! Walking and bicycle riding also require energy, but at least the energy you use is your own.

Giving up driving, however, is not the only way you can save gasoline. There are numerous factors which affect the gas mileage of your car. Some depend on characteristics of the car, such as its size, engine, and type of transmission. Once you've bought the car there's not much you can do about these. Other factors depend on your driving habits, and these are under your control. By observing a few simple tips given in this chapter, you may save several hundred gallons of gasoline a year. But first a word of caution: don't be disappointed if your car does not behave exactly as the following charts suggest. No two cars are alike, and no two cars behave the same way under the same conditions. The charts show you approximately what you can expect from your car. Use the charts as a guide, and experiment to find the conditions which give you and your car the best mileage. Every time you buy gasoline, fill your tank completely and make a note of the date, the mileage on the odometer, and the number of gallons you bought. Determine average mileage by dividing the difference in odometer readings by the number of gallons needed to fill the tank. A lower than usual mileage means higher gas consumption and signals trouble. Either your engine needs attention, or your driving habits have changed for the worse.

BUYING A NEW CAR

Automobiles are designed to sell. Over the years, the buying public has clamored for larger and larger cars, and the automakers have been obliging their customers. As a result, the average weight of standard size cars increased 30 percent during the past decade. The time has come to stop this trend.

Next time you're buying a car, don't choose it for its style alone, but look also at its weight. A lighter weight car will almost certainly cost less than a heavy one. Furthermore, lighter cars use less gasoline. A compact car weighing around 2800 pounds travels 17 miles on one gallon of fuel, while a standard size car gets only 11 miles and a luxury sedan only nine. On the other hand, a subcompact may average 24 miles per gallon — over twice as much as some of its bigger brothers.

TYPICAL WEIGHTS OF AUTOMOBILES

Sub compact	2200 lbs
Compact	2800 lbs
Intermediate	3500 lbs
Full size	4000 lbs
Luxury sedan	4600 lbs

A subcompact thus saves you nearly 500 gallons of gasoline a year — enough to drive a small car for 10,000 miles, or a larger one for 6,000 miles! If your family is too large for a subcompact, weigh the advantages of an intermediate car over a standard size one. In a year the intermediate will use 200 gallons less than the full size model, giving you an extra 3,000 miles of driving for free.

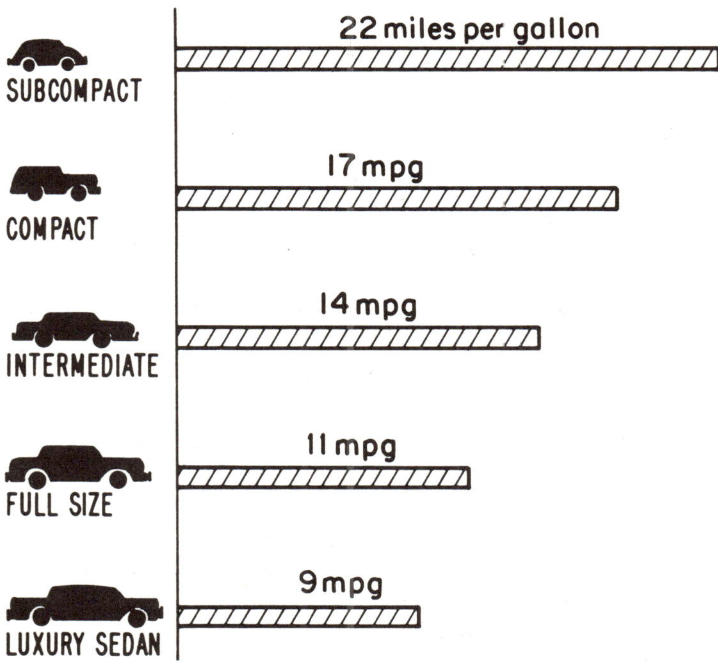

Typical Gasoline Mileages of Automobiles of Different Sizes.

Optional accessories make driving easier and more comfortable, but they also increase your gasoline consumption. The heaviest burdens on mileage are automatic transmissions and air conditioners. Each of these reduces your mileage by roughly 10 percent. Let's suppose you're an average driver driving an air-conditioned car with automatic transmission, and you drive 10,000 miles a year. If your car had a manual shift and no air conditioning it could go 2,000 miles more every year without consuming any additional gasoline. Incidentally, air conditioners waste gasoline even when they are not turned on, because of the extra power (and gasoline) needed to carry their added weight. You pay for the option of having air conditioning, then, whether you use it or not.

Alternators, which replace the old fashioned generators in many cars, also waste another 8 percent of your gasoline. Power steering costs an additional 1 percent. Now let's add up the increases in gasoline consumption due to all the various accessories: air conditioning 10 percent, automatic transmission 10 percent, alternator 8 percent, and power steering 1 percent. The total increasy in gasoline consumption is 29 percent. If you would be willing to forego the comforts provided by these devices, you would stretch your 10,000 miles per year to almost 13,000 miles per year without buying so much as an extra drop of gas.

POLLUTION CONTROL DEVICES

It is commonly believed that the pollution control devices on your new car increase gasoline consumption. The figures quoted generally show that fuel economy suffers 10 to 20 percent when control devices are added to a car. In fact, according to recent studies conducted by the Environmental Proection Agency, this does not hold for all cars. It is true that heavier cars pay a significant penalty in fuel economy when equipped with pollution control devices. A full size car with pollution control uses 13 to 18 percent more gasoline than one without controls. Interestingly enough, however, the gas consumption of small cars appears to improve with pollution controls. The gas mileage of an intermediate car increases by 1 or 2 percent with controls. The good sense of buying a small car is once more evident.

GASOLINE CONSUMPTION DUE TO ACCESSORIES

	Approximate Increase in Gasoline Consumption (city type driving)
Air conditioning	10%
Automatic transmission	10%
Alternator	8%
Power steering	1%
Total:	29%

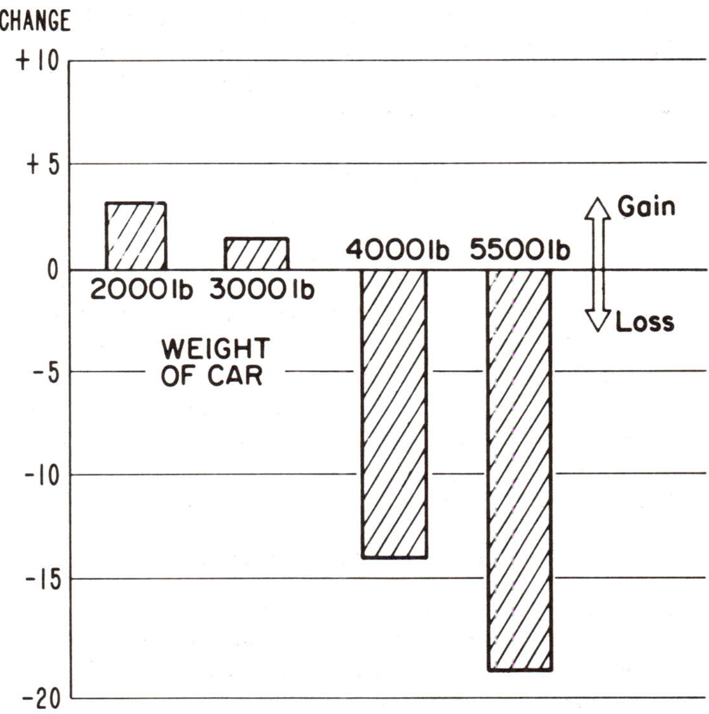

The Change in Mileage Due to Pollution Control Devices.

TIRES

Power, and therefore gasoline, is needed to overcome the friction between the tires and the road. The friction depends on the size of the car, on the tire pressure and on the type of tires on your car. For smaller cars, tire friction (and therefore fuel consumption) is less than for larger cars. This is another reason why smaller cars have better fuel economy than larger cars.

Tire pressure has a pronounced effect both on the fuel economy and on the life of the tire. Underinflated tires cause higher fuel consumption, and cause the tire to wear out quickly at the edges. If a tire is overinflated fuel economy may improve somewhat, but the tire will wear out in the center. And a rapidly worn-out tire costs you money. It also wastes energy, since energy is required in the production of new tires. For best results, then, always inflate the tires to the correct pressure. This pressure varies with the type of car, the tire, and with the driving conditions. Follow carefully the manufacturer's recommendations in your owner's manual.

The type of tire you are using also affects the gasoline consumption. Radial tires give about 3 percent better fuel economy than conventional bias ply tires.

One more word about tires: avoid spinning the wheels when you're stuck in snow, ice, sand, and so on. It wears the tires out fast, wastes gasoline, and just digs you deeper into a rut. Use snow tires in the winter for extra traction. If your car gets stuck in the snow "rock" it out by putting it alternately in forward and reverse. If you still cannot get moving ask for help.

GASOLINE

The type of gasoline you use affects the life, the performance and the economy of your engine. It is important, therefore, that you select the right mixture for your car. Some cars require high octane, premium gasoline while others may run satisfactorily on lower octane, regular gas. The octane requirements of cars vary even among cars of the same year and model. First, consult your owner's manual and buy the gasoline recommended by the manufacturer. Then experiment with different types of gasoline, and use the one that gives you the best mileage. After your car has been driven for several thousand miles you might start hearing a "knocking" or "pinging" sound from the engine, particularly when you press the accelerator or climb a hill. When this occurs switch to a higher octane gasoline.

TUNE UP AND MAINTENANCE

Every machine operates best when maintained properly. Automobiles are no exception. A properly maintained automobile requires less repair, lasts longer, and has higher gas mileage than one that has not seen regular service. After your car has been tuned, for instance, its mileage increases by about 6 percent, while a very poorly maintained or grossly mistuned car may use as much as 20% more gasoline than one that is in good working order.

For these reasons, keep your engine in tune and have the car serviced at regular intervals (the maintenance schedule for your car is described in the owner's manual). Some of the items that can give you problems and need periodic attention are: spark plugs, distributor, timing, air filter, carburetor, and pollution control valve. A dirty air filter or fouled spark plugs will really hurt your fuel economy. A maladjusted distributor or an incorrect timing also results in skyrocketing gas usage, while a maladjusted carburetor uses so much gasoline that you may have trouble getting to the nearest gas station. Wheel alignment is something else to keep an eye on. Badly aligned wheels are a drag on the engine and waste gasoline.

Engine temperature affects gas mileage, so make sure your thermostat is working properly. A faulty thermostat keeps the engine from warming up quickly in cold weather, and a cold engine uses more gasoline than a warm one. Your engine also needs a different type of oil in the winter, though you might try a multipurpose oil that's good both in winter and in summer. Your local service station can advise you what kind of oil is best for your car.

SPEED

As the automobile moves the surrounding air exerts a force on it. This force is called wind resistance or drag. Work needed to overcome this drag is supplied by the engine. As the wind resistance increases, so does the engine's work and so does your gas consumption. If wind resistance is low the engine's work becomes less and gasoline consumption decreases.

Wind resistance depends on the size and shape of the car and on the speed at which the car travels. Here the size refers to an area roughly equal to the product of the vehicle's height and width. The smaller this area the smaller the wind resistance, and the lower the fuel consumption. This is another reason why small cars are more economical to operate.

The shape of the car has also an important influence on the wind resistance. A streamlined car with gently sloping surfaces has less wind resistance than a car resembling a shoe box. Our cars are not designed to give the least wind resistance (and highest mileage) but are designed mostly with styling and sales in mind. Since there is nothing you can do about the shape of your car, there's no need to worry about it. But there is one factor contributing to wind resistance which you are in full control of: the speed of the car. More energy is needed to overcome the wind resistance at high speeds than at low speeds.

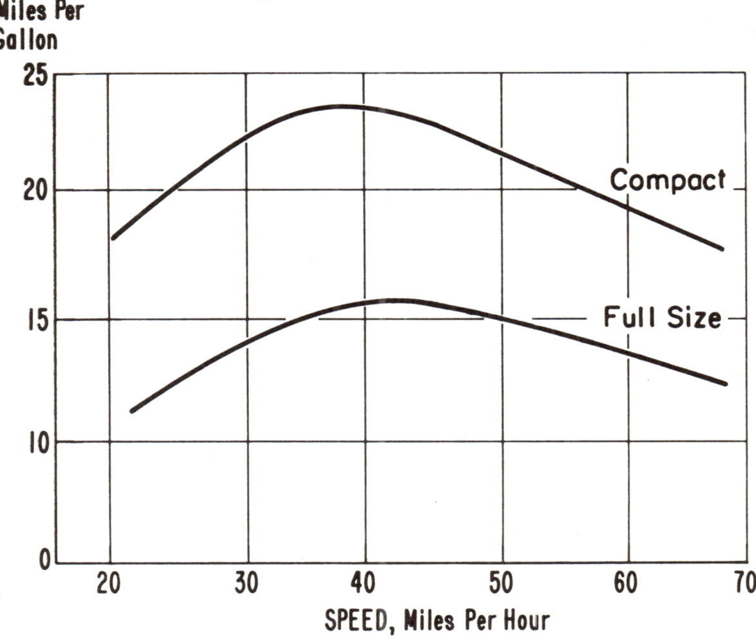

Fuel Economy at Constant Speeds.

This would suggest that the faster you drive the more power and consequently, the more gasoline you use. This is not entirely true, though. To understand fully the relationship between speed and fuel consumption we must consider two more factors besides wind resistance: rolling resistance and engine efficiency. Rolling resistance is caused by the friction of the tires, bearings and other moving parts. Engine efficiency indicates how much useful work is done by the engine compared to the energy (gasoline) put into it. The combined effects of wind resistance, rolling resistance and efficiency are such that the best mileage occurs when the car moves at a steady speed of about 40 miles per hour. At speeds above or below 40 mph gasoline consumption increases and the mileage becomes worse. The decrease in gasoline mileage is particularly evident at higher speeds.

Cruising at 60 instead of 70 miles per hour saves 15–20 percent on gasoline, and at 50 instead of 70 miles per hour saves 25 to 30 percent. Reducing the speed from 70 to 40 miles per hour results in a 50 percent greater mileage. To put it differently: if you drive a standard size car with a 20 gallon tank you can drive 200 miles at 70 mph, 240 miles at 60 mph, 280 miles at 50 mph or 300 miles at 40 mph on one tank of gas.

It would be wonderful if you could always drive at a steady 40 mph speed, but this of course is not possible. You must slow down, stop, and accelerate to keep up with the traffic. All these changes in speed cost you gasoline. Driving in the city, for example, uses 30 to 50 percent more gasoline than driving on the highway. You can save gasoline even in stop and go traffic, though, by changing speed as seldom and as gently as possible. If you must slow down or accelerate, do it gradually. Don't step on the brake suddenly, and don't put a "lead foot" on the accelerator. Moderate driving, as opposed to sudden stops and quick accelerations, can save you 15 percent on gasoline — enough to drive around the city one extra day a week.

STARTING A COLD ENGINE

Starting a cold engine may be a nerve wracking experience, but following a few simple rules can help you both in starting your engine and in saving gasoline. Your owner's manual may also give you specific instructions for starting your car.

First of all, try to keep your engine warm. Park the car in a garage and close the garage door. If you do not have a garage, park in a car port — even parking under a tree helps. If your garage is full of bicycles, tricycles, firewood and other paraphernalia, at least make enough room just to get the hood inside the garage. In extreme cold weather put a 60 watt light bulb under the hood. The electricity you waste is small in comparison with the trouble and gasoline it saves you.

Secondly, winterize your car before the cold weather sets in. Have your engine tuned and your battery charged. A fully charged battery is essential for an easy start, and you simply cannot start an engine on a run-down battery. Your service station can check the condition of your battery in seconds.

As we mentioned above, make sure you have the right type of oil for the winter.

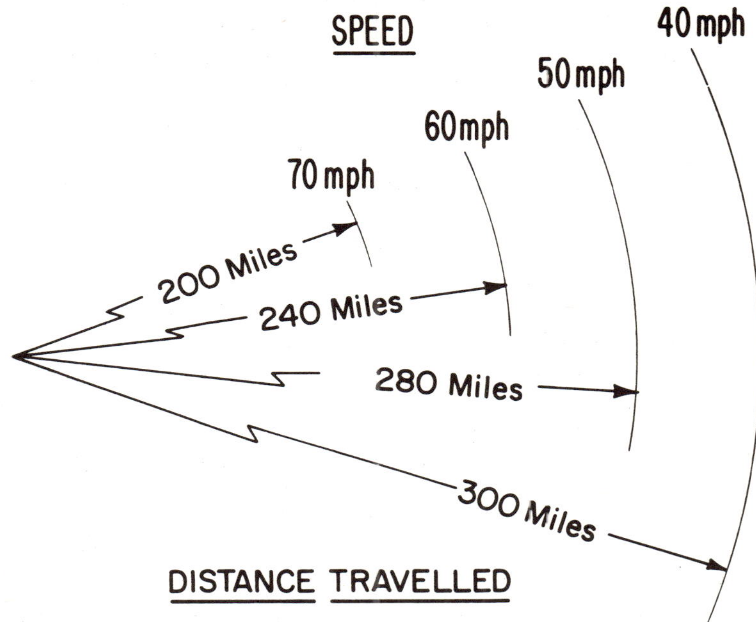

Range of a Standard Size Car on 20 Gallons of Gasoline.

A summer weight oil becomes thick and sticky in cold weather, and you have trouble starting no matter how good your battery is.

When you are ready to start your engine turn off all the accessories. Be sure the lights, radio, air conditioner, heater and defroster are all off before you turn the ignition key. If your car has a manual transmission depress the clutch: this eases the load on the engine. Finally, depress the accelerator once or twice and then turn on the ingition. If your engine won't start on the first try, wait a minute before you try again. Never crank the engine for more than 30 seconds. This wears down the battery and ruins any chance you may have for getting started.

Once the engine starts, warm it up fast because a warm engine uses less fuel than a cold one. But don't idle the engine and wait for it to warm up. Wait just long enough so the engine won't stall when you put it in gear, then get moving.

One more useful tip for driving in the winter: keep your tank full. A full tank adds weight to the rear end of the car and reduces wheel slipping, which wastes gas. A full tank also reduces the chance moisture will condense inside the tank. This helps prevent fuel line freeze-up.

STARTING A STALLED ENGINE

A stalled car is dangerous, inconvenient and embarassing. There you are sitting in your car right in the middle of an intersection while cars toot at you from every direction. In desperation you probably do just what you should not be doing: you keep cranking the engine in the hope that, by some miracle, it will fire up again. The cardinal rule of starting a stalled engine is: never crank it too long. Cranking will not only not start your engine, but most likely will prolong your agony.

When your engine stalls, first determine the cause for the stall. Glance at the fuel gauge. If it is on empty, there is nothing doing: raise your hood and start walking to the nearest service station. If you are not out of gas, your troubles are likely to be caused by vapor lock, carburetor icing, or a flooded engine.

Vapor lock usually occurs on warm spring days before your service station has switched over to summer type gasoline. Vapor lock is caused by vapor formed in the fuel system due to the combined effects of high air temperature, engine heat, and a gas mixture designed for winter. The vapor may prevent the gasoline from reaching the carburetor and the engine. It is very difficult to restart an engine stalled by vapor lock until the vapor disappears, but your best bet is to wait until the engine cools down. If you have cold water handy, wrap a wet cloth around the fuel line leading to the carburetor. This may help dissolve the vapor.

Flooding generally occurs on hot summer days when you slow down or stop. While the car is moving, a stream of air rushes across the grille, cooling the engine and the carburetor. When the car stops this air flow also stops, allowing the engine and the carburetor to heat up. If the temperature becomes too high, the gasoline starts boiling in the carburetor and spills over into the manifold leading to the cylinders, thereby "flooding" the engine. When this happens wait a few minutes to allow the engine to cool down. Then put the gear in neutral, depress the accelerator all the way to the floor, and turn on the ignition. The engine should start in 30 to 40 seconds. If the engine does not start in this time wait 5 minutes, and try again.

Carburetor icing is most likely to occur on damp rainy days, when the temperature is between 30 and 40°F. It is caused by the moisture in the air which freezes out in the carburetor. The ice which forms on the throttle interferes with the flow of the gasoline-air mixture, and you can't restart the engine until this ice layer has melted. Don't try to restart the engine immediately because this will just compound the problem. Wait a while until the heat from the engine melts the ice, and then try your ignition key. If it's too cold out the ice may not melt unless the car is towed to a heated garage.

IDLING

Fuel economy, or gasoline mileage, is the distance traveled in miles per gallon of gasoline. When the engine is idling the car is using up gasoline without going anywhere, so its fuel economy is zero. Therefore, don't idle your engine for long periods. Shut off the engine when you are caught in a really heavy traffic jam or when you are waiting for someone. An idling engine consumes a gallon of gasoline an hour while only a spoonful of gasoline is needed to restart the engine.

AIR CONDITIONING AND HEATING

Power for the air conditioner in your car is provided by the engine. Air conditioning may increase the gasoline consumption by 5, 10 or even 15 percent, depending upon your car and the weather. You save gasoline, if you turn the unit off. You also save, but not as much, if you turn the temperature control to a higher setting. Tinted windows and parking in the shade also help to keep your car cool.

Heating the car, on the other hand, is almost free. The heat used for warming the car comes from the waste heat of the engine. Since this heat would be discarded anyway, you don't pay for it. The fan forcing the heat into the car uses some power but not very much: it increases fuel consumption by a mere 1 percent.

SHORT TRIPS

It is tempting to jump in the car and drive to the corner drugstore, or drive down the street to see a friend. Such short trips, unfortunately, are real gasoline guzzlers. It takes 10 to 15 miles to warm up your engine, and until your engine is warmed up your gas mileage is poor. For instance, one gallon of gas will take you 5 miles when your engine is cold. Once the engine is warmed up — that is, once you've already gone 10 or 15 miles with low mileage — a gallon of gas will take you 15 miles. Because your fuel economy is so poor on short trips, the thing to do is to avoid them. Walk to nearby locations, or plan your day so you can do all your errands at once.

WEATHER CONDITIONS

The weather has a significant effect on the gasoline consumption of your car. In cold weather, fuel economy suffers. In the first place, in cold weather it takes longer for the engine to warm up, and a cold engine uses 20–30 percent more gasoline than a warm one. The weather also affects the operation of an already warm engine. You get better mileage in the summer, because for each 10°F increase in temperature your car uses 2 percent less gasoline. Thus at 80°F you need 10 percent less gasoline than at 30°F.

Gas mileage is also different at different altitudes. At high altitudes the carburetor and parts of the ignition system operate differently than at lower elevations. To save gas, readjust your carburetor and ignition if you move to an area that is at a markedly different altitude from where you lived before.

Finally, a word on the effects of wind. A tailwind helps you move ahead while a headwind hinders your progress. Unfortunately, you have no more control over the direction of the wind than you have over temperature and altitude. The most you can do about the weather is try to save your big trips for the summer.

ROAD CONDITIONS

You have certainly experienced the difficulty of driving on a bad road or on a

Trip Length

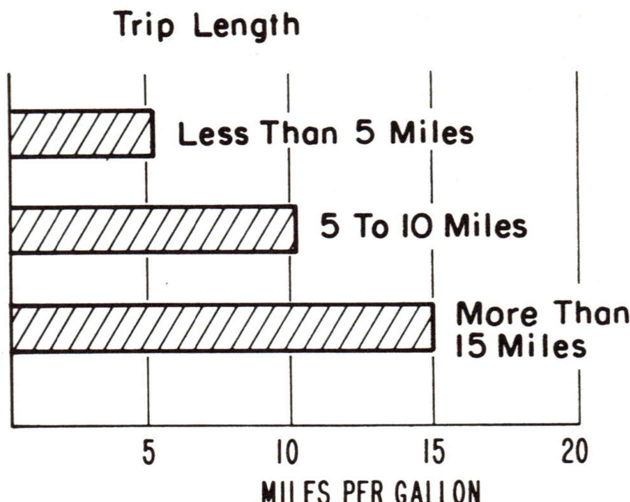

Gasoline Mileage for Trips of Various Lengths
(Starting with a Cold Engine)

snowy street. You could almost feel the engine laboring as it was trying to push the car forward. It won't come as a surprise, therefore, that more gasoline is used up on a poor road than on a good one. It may surprise you, however, just how much difference the road can make in fuel economy. Driving on patchy asphalt road can cost you 15 percent more gasoline than driving on a nice smooth street or highway. Driving on a dirt or gravel road may cut your gasoline mileage by 25 to 30 percent, and on sand or snow your gasoline consumption is nearly twice as much as on a dry surface. It pays therefore, to avoid bad roads.

Winding and hilly roads also require extra fuel. On a winding road the frequent changes in speed cut into your fuel economy. Going uphill, the engine must produce extra power to lift the weight of the car, so your gas consumption may increase by as much as 50 percent. Following the principle that "everything that goes up must come down", you may figure on gaining this back when you drive down hill, but such is not the case. You would indeed save some if you would shut off the engine (or put the gear in neutral) and let the car coast. These practices, however, are extremely dangerous. Don't use them for the sake of saving some gasoline.

TRAILERS

Trailers are a marvellous invention. You can load all your belongings into them and move to a new home. Or you can pack up your camping gear and take your family for a nice vacation. But you pay energy for this convenience. It takes power

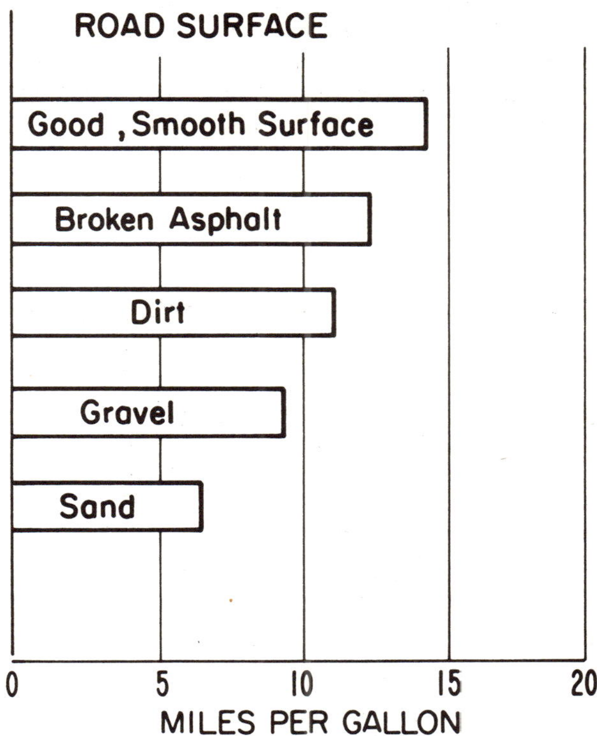

Fuel Economy on Various Road Surfaces.

to pull the trailer behind your car and your engine must provide this power. The result: increased gasoline consumption. The amount of extra gasoline you need to tow a trailer depends, of course, on the trailer and on the speed at which you travel. The weight of the trailer is a very important factor. As a rough rule, it takes twice as much gasoline to pull a 2000 pound trailer as it does to pull a 1000 pound one, three times as much gasoline to pull a 3000 pound trailer as it does to pull a 1000 pound one, and so on.

The wind resistance caused by the trailer also adds to your fuel consumption. A trailer with a low profile offers less wind resistance than a tall, box-like trailer; hence, it takes less gasoline to pull a low trailer than it does to pull a tall one. If you are buying or renting a trailer for camping look for one that is small on the road and can be opened, if necessary, at your camp site.

When you tow a trailer your speed makes a big difference in gasoline use. The faster you go the bigger the wind resistance and the greater your gasoline consumption. At 60 miles per hour you use up nearly twice as much gasoline as at 40 miles per hour. Thus, by traveling at 40 instead of 60 miles per hour you "pay" only for half the distance and save enough gasoline to drive the other half "free".

Underinflated tires on the trailer are a drag on the engine and waste fuel. Make sure that the tires are inflated to the correct pressure. Check also the connection between the trailer and the car. A badly hooked up trailer sways behind your car, making driving difficult and unpleasant. This swaying or "tracking" also adds to the load on your engine and to your gasoline cost.

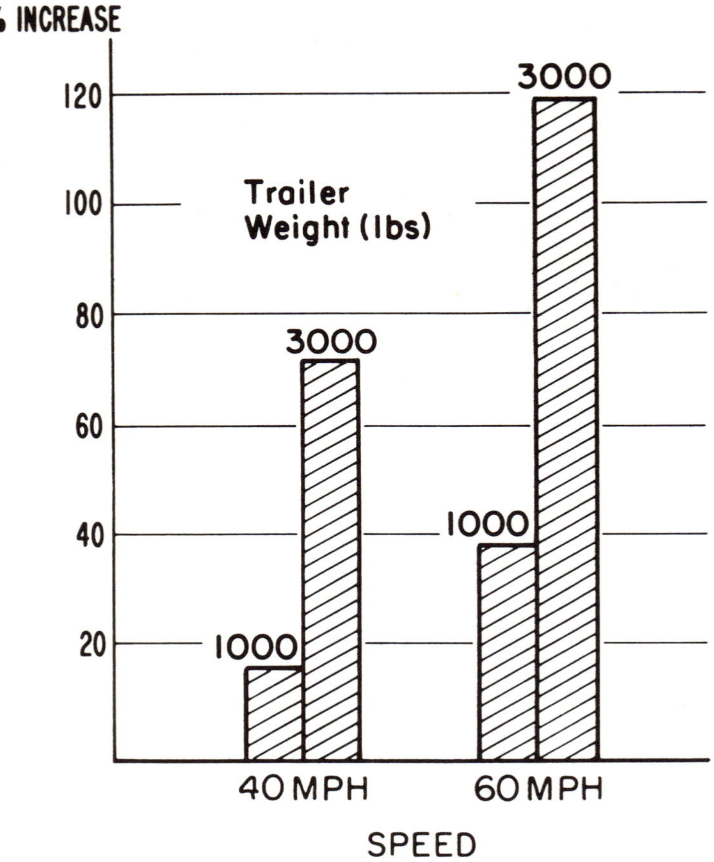

Increase in Gasoline Consumption of a Passenger Car Towing a Trailer.

LUGGAGE RACKS

If you're like a lot of people, you're apt to leave behind the things you need most when you go on a vacation. Expecting sunny days, you leave your raincoat at home — and find that it rains all through the trip. Hoping to go swimming, you carry along your bathing suit, but the weather turns cold, and you wish you had your heavy sweaters instead. The next time, then, you decide to take everything you can on the trip. Of course, all the suitcases won't fit in the trunk, but you put the extra luggage into a luggage rack. You also hitch the skis onto the car and, just in case, fasten the bicycles on the outside of the trunk. You are now assured that you have everything you need with you — everything, except enough gasoline. All that luggage outside the car increases the wind resistance and cuts your gasoline mileage, so that for your excess luggage you are paying a 1 or 2 mile per gallon penalty. If you can help it, then, don't carry anything outside the car. Load everything inside the trunk or into the back of the station wagon. If you must carry anything outside the car, drive slowly. At lower speeds the wind resistance caused by the luggage is smaller and you are wasting less gasoline.

CAR POOLS

There is nothing more convenient than driving your own car to work. You can start in the morning whenever you are ready, skip out at noon to run an errand, and leave work a few minutes early to catch your daughter's birthday party. Driving alone is certainly convenient — and just as certainly wasteful. Two, three or four people teamed up in a car use a fraction of the gasoline which is used up when everyone travels separately.

The car pool is one obvious answer to the gasoline shortage. Even if we allow for the fact that a little extra driving is necessary to pick up and drop off the various members of the pool, the pool can save you truly enormous amounts of gasoline. Suppose, for instance, you have to drive 10 miles to work, or a total distance of 20 miles for the round trip. You need approximately 2 gallons of gasoline a day, or roughly 50 gallons a month. If you share the driving with one other person you use only about 26 gallons monthly (allowing for the extra driving to pick up and drop off your friend). Thus, you save 24 gallons a month, or about 300 gallons a year — enough gasoline to drive 3000 to 3500 miles! If there are three or four of you in the car pool you save even more. With three in the car you use only 18 gallons of gasoline each month; with four it costs you only about 14 gallons, and you save 430 gallons in a year. On this much gasoline you could drive from New York to San Francisco, and back again.

A car pool has other advantages besides saving gasoline and reducing wear and tear on your car: you have less worry about driving and parking. Car pools also mean fewer cars on the road, easier traffic and less air pollution.

PUBLIC TRANSPORTATION

Public transportation almost always requires less energy than your car. Mass transit in the city uses only about half as much fuel per passenger as the private automobile, and it also saves the headaches and the cost of parking. Rapid transit systems will also get you there faster than your car will.

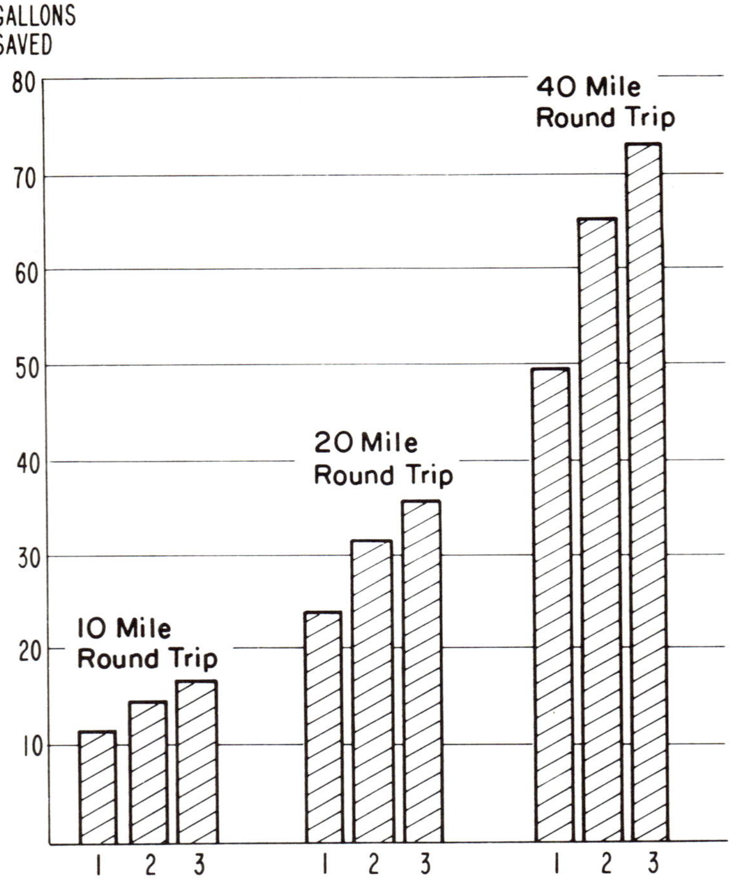

NUMBER OF PASSENGERS (EXCLUDING DRIVER)

Car Pools
Gasoline Saved by each Participant in One Month.

For long distance travel both the train and the bus are more economical than going by car. If you are really concerned by the fuel shortage you should take the bus. Long distance buses burn only half the fuel per passenger that cars and trains use.

Public transportation not only saves energy, but it is also generally cheaper than private travel. For long distances it even costs less to go by jet plane than by car. The cost of travel in terms of both energy and money depend, of course, on the number of passengers in your car. The car is very uneconomical when it carries only one person, the driver. Its economy improves as the number of riders increases.

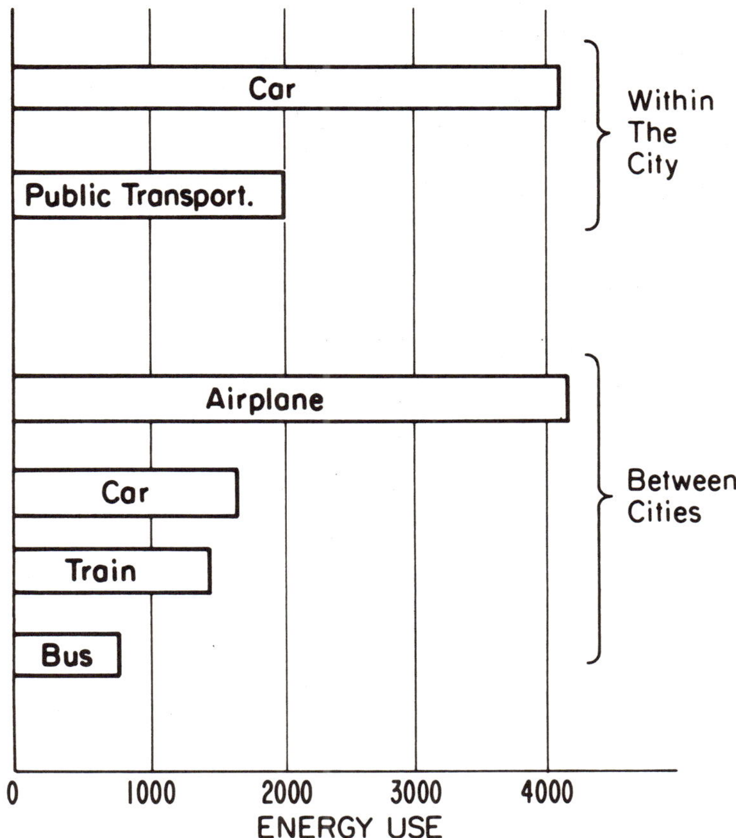

Energy Requirements of Different Modes of Transportation.
(Btu per Passenger Mile)

With three riders — the driver and two passengers — the car becomes competitive in cost and fuel consumption with other modes of transportation. However, from the point of view of air pollution, public transport is still preferable. A bus with 60 passengers pollutes much less than twenty cars with 3 riders each.

PLANNING THE TRIP

As we have seen there are many conditions under which your gasoline mileage suffers. You pay a penalty for driving in cold weather, on bad roads or in heavy traffic. You burn less gasoline and money if you prepare your trip carefully to avoid these conditions, and follow a few simple rules:

First, before you go on a long trip have your car serviced. Your mechanic might catch some problems which can cause you trouble and expense on the road, and a car that is tuned and adjusted also needs less gasoline. Secondly, try to make longer trips in the summer when the weather is good and the air is warm. You are less likely to run into bad weather and the extra daylight makes driving easier. Furthermore, in warm weather your gasoline consumption is less than in cold weather.

Select your route carefully. Avoid bad road surfaces, hilly and winding roads, and weigh the advantages of detours, which may save you fuel and time. Try to avoid heavy traffic. Time your travel so that you don't have to drive near or across cities in peak traffic hours. If you must be near a city at 5 PM consider alternate routes around the city rather than driving through downtown. The time and gas you save may be worth the extra few miles. If all else fails, take your rest stops at peak traffic times, and wait till the traffic subsides.

Last but not least, allow enough time for the trip. Leave in time, and drive leisurely at a comfortable, safe and fuel-saving speed. High speeds are dangerous and uneconomical.

MOTORCYCLES

If you are reluctant to walk, unwilling to ride a bicycle, or unable to take the public transport, a motorcycle may be the answer for you. It will get you there quickly and cheaply, although it does have its disadvantages. Its chief drawback is that it provides little protection from the elements and is more dangerous than a car.

As a gasoline saver, though, the motorcycle is great. A small 10 horsepower motorcycle gets 80 to 100 miles per gallon, and you can ride one around for a whole week on one or two gallons of gasoline. Not every motorcycle will give such good mileage: a large 50 hp bike uses as much gasoline as an automobile. If you want a motorcycle for conserving fuel, then buy one that has a low horsepower. Four stroke engines are more economical than two stroke ones and will add another 5 to 15 percent to your mileage.

Whatever else you do on your motorcycle, don't speed. High speeds increase your fuel consumption by about 5 to 10 percent for every 10 mph; more importantly, they increase your chances of being seriously injured. At low speeds you are much safer and your fuel economy is better. Another thing to avoid is carrying passengers on your bike: it decreases mileage and increases risk.

GASOLINE CONSUMPTION OF MOTORCYCLES

When the Cruising Speed Changes From	Gasoline Consumption Increases by About
40 to 50 mph	12%
40 to 60 mph	17%
40 to 70 mph	25%

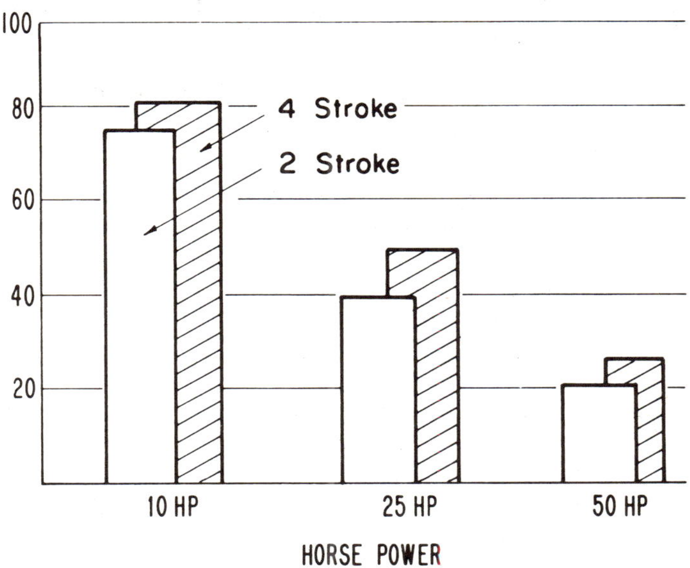

Typical Gasoline Mileage of Motorcycles in City Traffic.

A word about those knobby tires. They may be fashionable but they are unsafe and uneconomical. By all means use regular tires and make sure they are inflated to the proper pressure.

MOTOR HOMES AND CAMPERS

In the summer it often seems as if motor homes and campers have taken over the highways. And for good reason. You can load your family with all your vacation gear into one and off you go. You don't have to limit your luggage, you're not tied to train, bus, or plane schedules, and you can stop whenever you like. No worries about motel vacancies or expensive restaurants either. You can camp at any campsite and if you don't like it you can go on.

So much for the good things about campers. Now let's see the other side of the story. Motor homes and campers, as you probably have suspected, use a lot of gasoline. Since they are heavier than passenger cars they need extra power, and (just as in the case of cars) the heavier the camper, the more gasoline it consumes. The box-like shape of the campers makes them comfortable inside but also increases their wind resistance, adding further to gasoline consumption. Because of their increased wind resistance campers and motor homes are more sensitive to changes in speed than passenger cars. Increasing the speed of a camper from 40 to 60 miles an hour results in a 30 percent increase in gasoline consumption, and traveling at 70 mph instead of 40 mph actually cuts mileage in half. It is particularly important, then, to drive your camper or motor home at a moderate speed.

GASOLINE CONSUMPTION OF MOTOR HOMES

An Increase in Cruising Speed From	Results in an Increase in Gasoline Consumption of About
40 to 50 mph	20%
40 to 60 mph	30%
40 to 70 mph	45%

Campers and motor homes are the same type of machinery as automobiles — they are just bigger. Therefore, the same fuel saving rules apply to them as to passenger cars. Service them regularly and keep them well tuned and maintained. Drive them carefully, avoiding drastic stops and fast acceleration. Drive in good weather and on good roads. Avoid heavy traffic and don't idle the engine. And most of all — drive slowly.

CHECKLIST FOR THE AUTOMOBILE

The new car buyer

> Could you do without the car?
> Are you buying a car as small as possible?
> Do you need a car with automatic transmission?
> Must you have air conditioning in your car?

Tires

> Do you have the right type of tires for your car?
> Can you replace the conventional tires with radials?
> Are your tires inflated to the proper pressure?
> Are you using snow tires in the winter?

Tune up and maintenance

> Is your car serviced and tuned at regular intervals?
> Are the spark plugs and air filter clean?
> Is the ignition system working correctly?
> Is the carburetor adjusted properly?
> Have you checked the pollution control valve?
> Are the wheels aligned?

Is the thermostat working properly?
Are you using the right type of oil?
Is your battery charged?
Is your car winterized?
Is the gas tank full?

Gasoline

Are you using gasoline with the right octane rating?
Are you switching to a higher octane gasoline when the engine begins knocking?

Starting a cold engine

Do you keep the car in the garage in the winter?
Have you turned off the accessories before starting?
Have you depressed the accelerator once or twice before turning the ignition key?
Are you cranking the engine too long?
Are you waiting for the car to warm up instead of getting under way?

Starting a stalled engine

Is your car out of gasoline?
Have you determined the cause of the stall (vapor lock, carburetor icing or flooded engine)?
Have you waited long enough before trying to restart the engine?
Is the gear in neutral?
Are you cranking the engine unnecessarily?
Have you floored the accelerator when trying to restart a flooded engine?

Driving

Are you driving too much?
Could you take the public transport?
Could you join a car pool?
Are you driving too fast?
Are you changing speed too often?
Are you stopping and accelerating too frequently and too abruptly?
Do you let the engine idle too long?
Are you spinning the wheels on slippery surfaces?
Do you turn on the air conditioning unnecessarily?
Do you carry luggage outside the car?

Planning the trip

Have you planned your trip carefully?

Are you avoiding short trips?
Are you driving in warm weather, whenever possible?
Have you selected good roads for your trip?
Are you avoiding hilly and winding roads?
Are you avoiding heavy traffic?
Have you allowed enough time for your trip?

Trailers

Is your trailer as light and small as possible?
Is it hitched on correctly?
Are the tires inflated properly?
Are you driving at a reduced speed?

Motor homes and campers

Is your motor home or camper too big?
Are you using it too much?
Are you driving slowly enough?
Are you following the same good maintenance procedures and driving habits as for your regular car?

Motorcycles

Are you riding a motorcycle with a low horsepower engine?
Does it have a four stroke engine instead of a two stroke one?
Does your motorcycle have regular tires?
Are the tires inflated to the recommended pressure?
Are you driving slowly?
Are you carrying an extra passenger unnecessarily?

CHAPTER IX

BOATS, SNOWMOBILES AND SMALL AIRPLANES

    Boats, snowmobiles, and small planes can burn a lot of gasoline and burn it fast. During an energy shortage, then, they should be used prudently. For some trips their use is justified, but on other occasions they might best be left at home. Whether or not a trip is warranted, of course, is for you to decide. If you forego the use of your boat, plane, or snowmobile for the duration of the energy crisis you don't have to worry about its fuel consumption. If, on the other hand, you decide to go for a ride, then you should at least observe a few energy-saving rules. These rules can save you gallons of fuel, give you extra miles of travel, and often improve upon the performance and safety of your craft.

BOATS

The amount of energy your boat consumes depends largely upon its size. You may own a large ocean going cabin cruiser, or you may have a small run-about which is just big enough for fishing in your local lake. Naturally, the fuel consumption of the two boats will be quite different. In fact, the larger boat may need fifty to a hundred times as much fuel as the smaller one.

For example, a small boat with a 2 to 5 hp outboard motor cruises along on only 0.1 to 0.2 gallons of gasoline per hour. Under similar conditions a boat with a 25 hp motor will use about 2.5 gallons per hour, and a larger boat will guzzle even more. A boat with a 115 hp outboard motor will need 6 gallons per hour. As a rule of thumb, fuel consumption under cruising conditions increases by about 1/2 gallon per hour for each 10 hp increase in the size of the motor. Thus you're not doing much to the nation's fuel supply if you go fishing in your 5 hp dinghy, but you're a real contributor to the energy crisis when you travel for hours in your 100 hp cabin cruiser.

Like automobiles, boats use more gasoline as they go faster, but the increase is even more dramatic than it is for automobiles. A car uses 50 percent more gasoline when traveling at 70 mph instead of 40 mph; in comparison, a boat burns ten times more gasoline at its top speed than it does when it's moving slowly. At "trolling" speeds almost any boat with an outboard motor uses less than 1 gallon of gasoline per hour. As its speed increases the boat starts making waves, requiring more fuel, so that at "cruising" speeds it's using 3 to 6 times as much gasoline per hour as it does when trolling. At top speed gasoline consumption (in gallons per hour) is nearly twice as high as at cruising speed — that is, 6 to 12 times what the boat burns per hour when trolling.

One thing to remember: fuel consumption (in gallons per hour) always increases with speed, even when the boat is planeing. But mileage (in miles per gallon) may improve at higher speeds because the boat covers a greater distance in the same amount of time. If you want to run your boat for a certain period of time, and you don't really care where you get during that time, then low speeds are best. But if your main concern is getting from one place to another, then use the throttle setting which gives you the best mileage.

The load you carry with your boat also affects its fuel economy. The size of the load, of course, makes more of a difference for a light boat than for a heavy one. In a small boat one or two extra passengers may cut mileage by 15 to 20 percent. For larger boats performance is not altered noticably by a moderate increase in the load.

Of course fuel consumption increases when your boat is used for towing. The effect is noticeable particularly at high speeds, but it's hard to predict exactly what the increase in fuel consumption will be. That depends on the conditions. For example, a water skier may add 10 percent to your gasoline usage, but towing the skier does not require much extra fuel in itself. Most of the extra gasoline is wasted by the frequent starts and stops.

APPROXIMATE FUEL CONSUMPTION OF BOATS AT VARIOUS SPEEDS

Horsepower	Gasoline Consumption, Gallons Per Hour		
	Trolling	Cruising	Top Speed
115	0.9	6	10
100	0.8	5	9
75	0.7	4	7
50	0.6	2.5	5
25	0.5	1.5	4
10	0.3	0.5	2
5	0.2	0.25	0.5
2	0.1	0.1	0.1

The weather can also raise havoc with your fuel economy. While your boat may need little gasoline to travel on calm waters, it will go much slower, and use more gasoline, in choppy or wavy water. Just how much extra gasoline you need in bad weather depends on the direction of the wind and waves and on the size of the waves relative to the length of your boat. In rough weather mileage can be dismally low. It's best to stay in port during bad weather, but if you must go out, drive slowly. At reduced speeds you save fuel and avoid possible damage to yourself and your boat.

Your propeller is one of the most important components of your boat, and for maximum fuel economy it is extremely important that it is properly matched to your motor and your boat. The wrong propeller reduces your motor's performance, keeping you from reaching top speed and also hurting your fuel economy. Be sure, therefore, that your propeller is suitable for your boat. The standard propeller which came with the boat may give you adequate service, but still not provide the best performance. Check with your dealer if you need help in selecting a propeller.

Your boat and motor must be maintained properly if you want to conserve gas. Tune and service the engine regularly, paying special attention to the spark plugs. They may need frequent cleaning or changing. If you are not planning to use your boat for several months, drain the fuel out of the entire system, including the fuel tank, the fuel lines, and the carburetor. Fuel left in the system deteriorates (particularly in warm weather) and deposits harmful gummy substances which may make your engine inoperative and require an extensive — and expensive — cleaning later on. Periodically clean your hull too. Weeds, barnacles and other deposits cut your speed and increase fuel consumption.

SNOWMOBILES

Not too long ago snowmobiles were a novelty. There were only a few around, and most of these were powered by small, lawnmower-type engines. Today, thousands of snowmobiles are sold every year, many with engines large enough to power an automobile. And along with the increase in engine size has come an increase in gasoline consumption. If you use your snowmobile strictly for pleasure, the best idea is to leave them parked during the energy crisis. If you must ride one, though, it'll help if you choose a low-horsepower model. A four-stroke engine will also give you better mileage than a two-stroke one.

Every type of vehicle consumes more gasoline at high speeds, and snowmobiles are no exception. To save fuel, therefore, you should drive at moderate speeds, preferably below 25 to 30 miles per hour. At higher speeds your fuel consumption goes up and you run the risk of injury to yourself and costly damage to your snowmobile.

When you stop your snowmobile, of course, always turn off the engine. Idling wastes fuel. Idling may also cause the engine to overheat and to die on you, and then it can be difficult to restart.

Weather and snow conditions change the fuel economy of snowmobiles significantly. In cold weather your fuel economy becomes worse: a 30°F drop in temperature can increase gasoline consumption by 5 to 7 percent. On wet snow fuel economy also suffers. To provide the power needed to overcome the extra friction, your engine may use 30 to 40 percent more gasoline on wet snow than it does on powder snow.

The load carried by your snowmobile is one more factor in your gasoline mileage. Depending on the snowmobile, an extra passenger may waste 5 to 10 percent gasoline. Towing has an even more important effect on fuel economy, sometimes reducing mileage by as much as 70 percent. The actual loss in performance depends upon the snowmobile, the surface conditions, and the object being towed.

Snowmobiles must be maintained in excellent condition. Poorly serviced engines waste fuel. They can also break down when you need them most, and when you're all alone in sub-zero weather that means trouble. Follow the manufacturer's recommendations in storing your snowmobile. Be particularly sure that all gasoline is cleaned out of the fuel system when the snowmobile is unused for long periods of time.

SMALL AIRPLANES

You might wonder why anybody should be concerned at all with private planes. You may dismiss them as a rich man's toy and a complete waste of fuel. In fact, flying is wasteful when it is used merely for pleasure (and, incidentally, so is driving a car, a boat, a camper, a snowmobile and any other vehicle). But the situation changes when small planes are used for serious traveling. Then planes, even small ones, become competititve with automobiles in terms of mileage. A single engine plane, for instance, averages 10 miles per gallon — nothing to brag about, but not bad in comparison with the gasoline consumption of many recent model cars.

TYPICAL FUEL CONSUMPTION OF SINGLE ENGINE AIRPLANES

		Fuel Consumption	
% Power	Air Speed (MPH)	Gallons per Hour	Miles per Gallon
75	160	14.5	10.5
65	150	13	11
55	140	12	11.2

So while small planes may require more fuel per passenger than public transport, they do offer a reasonable alternative to cars. Whether you're able to afford that alternative is another question.

CHECKLIST FOR BOATS AND SNOWMOBILES

Boats

Is the trip necessary?
Is the weather suitable and safe for boating?
Are you carrying an unnecessary load?
Are you going too fast?
Are you towing unnecessarily?
Is your boat equipped with the proper propeller?
Is your boat maintained and serviced well?
Is your boat stored properly in the off season?

Snowmobiles

Is the ride necessary?
Are the weather and snow conditions suitable and safe for snowmobiling?
Are you driving slowly?
Are you carrying an extra load?
Are you towing an unnecessary object?
Do you turn off the engine when stopping?
Is your snowmobile maintained in top condition?
Is your snowmobile stored appropriately?

APPENDIX I

ABBREVIATIONS

British thermal unit	Btu
Gallon per hour	gph
Horsepower	hp
Kilowatt	kw
Kilowatt-hour	kwh
Miles per gallon	mpg
Miles per hour	mph
Watt	w

HOW TO CONVERT VARIOUS ENERGY UNITS

Watt	divided by	1000	gives	kilowatt
Kilowatt	multiplied by	number of hours	gives	kilowatt-hour
Kilowatt-hour	multiplied by	3413	gives	Btu
BTU	multiplied by	3/10,000	gives	kilowatt-hour
Horsepower	multiplied by	745	gives	watt

APPROXIMATE HEATING VALUES OF VARIOUS FUELS

Natural gas	1,000	Btu per cubic foot
Heating oil	145,000	Btu per gallon
Coal	20,000,000	Btu per ton
Gasoline	125,000	Btu per gallon
Diesel fuel	140,000	Btu per gallon

ACKNOWLEDGMENTS

We gratefully acknowledge the technical information and suggestions which we received from many of our colleagues, both at The University of Michigan and elsewhere. We especially wish to thank Mr. A. Feeney for editing the manuscript, Mrs. S. Marty for her help with the figures, and Miss M. Hudkins for the countless times she typed and retyped the manuscript. We also want to thank our wives for their patience and encouragement, which helped make the book possible.